Stone Canyons

OF THE COLORADO PLATEAU

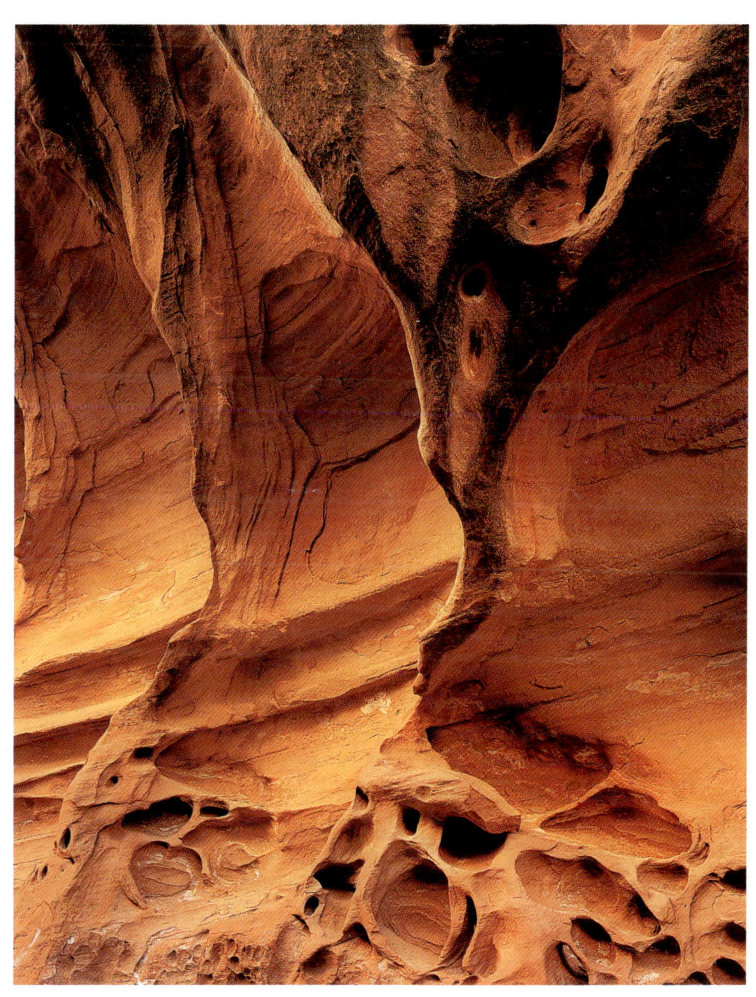

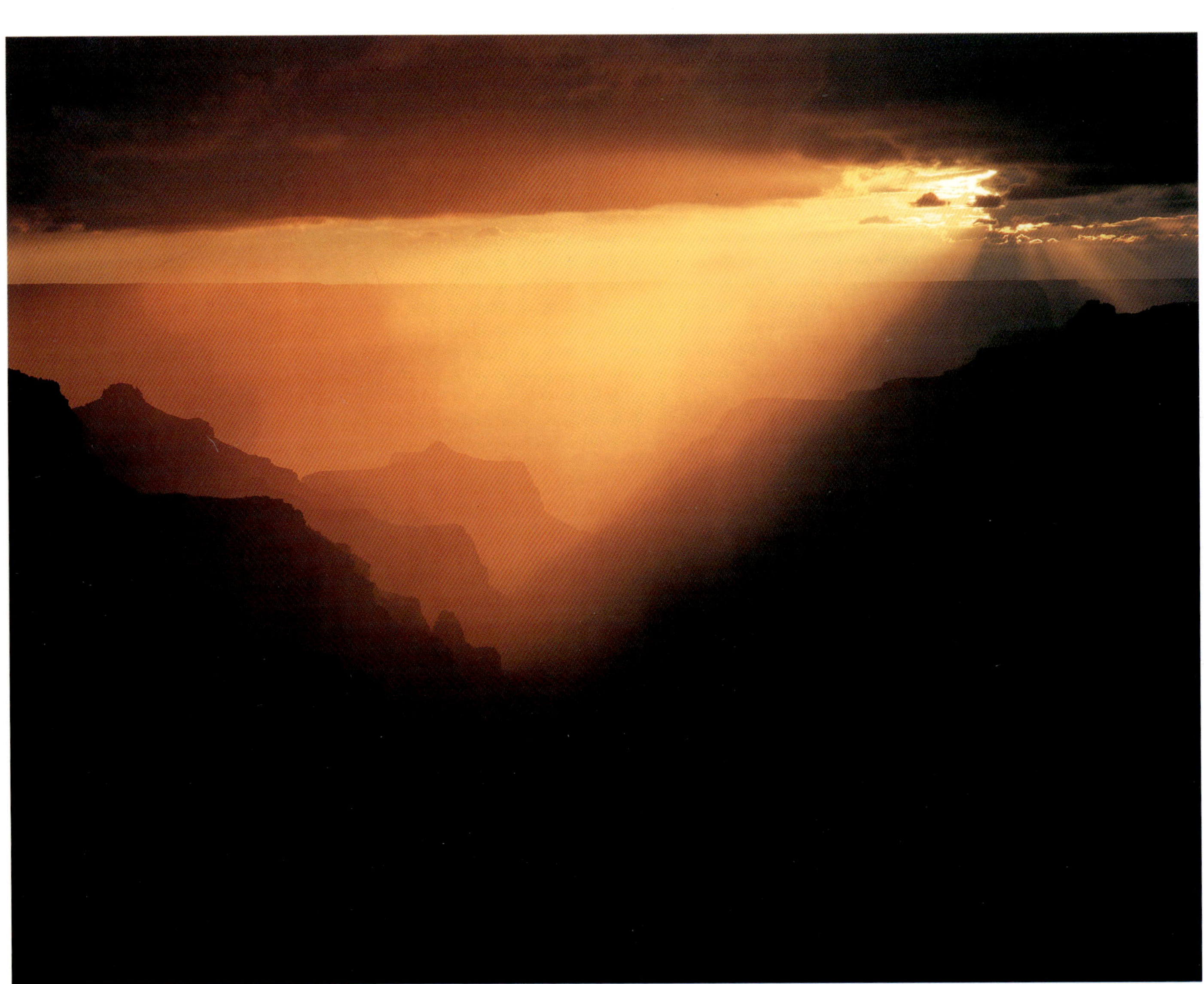

Stone Canyons

OF THE COLORADO PLATEAU

Photographs by Jack Dykinga/Text by Charles Bowden

Foreword by Robert Redford

HARRY N. ABRAMS, INC., PUBLISHERS

(page 1)

Narrow formations in the Wingate sandstone walls of Little Death Hollow. (Escalante, Wilderness Study Area, BLM, Arizona)

(page 3)

Sunset lights up a passing rainstorm and the ridges below Cape Royal. (North Rim, Grand Canyon National Park, Arizona)

For Margaret — Jack Dykinga

Editor: Robert Morton
Designer: Joan Lockhart

Library of Congress Cataloging-in-Publication Data

Dykinga, Jack W.
Stone canyons of the Colorado Plateau / photographs by Jack W. Dykinga : text by Charles Bowden.
p. cm.
ISBN 0-8109-4468-5 (cloth) / ISBN 0-8109-2992-9 (pbk.)
1. Natural history—Colorado Plateau. 2. Natural history—Colorado Plateau—Pictorial works. 3. Canyons—Colorado Plateau. 4. Canyons—Colorado Plateau—Pictorial works. 5. Colorado Plateau—History. I. Bowden, Charles. II. Title.
QH104.5.S6D95 1996
917'.91'3—dc20 95-40455

Paperback edition published in 2001 by Harry N. Abrams, Incorporated, New York

Clothbound edition published in 1996 by Harry N. Abrams, Inc.

Printed and bound in Hong Kong

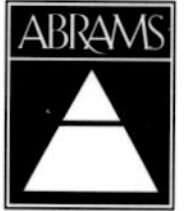

Harry N. Abrams, Inc.
100 Fifth Avenue
New York, N.Y. 10011
www.abramsbooks.com

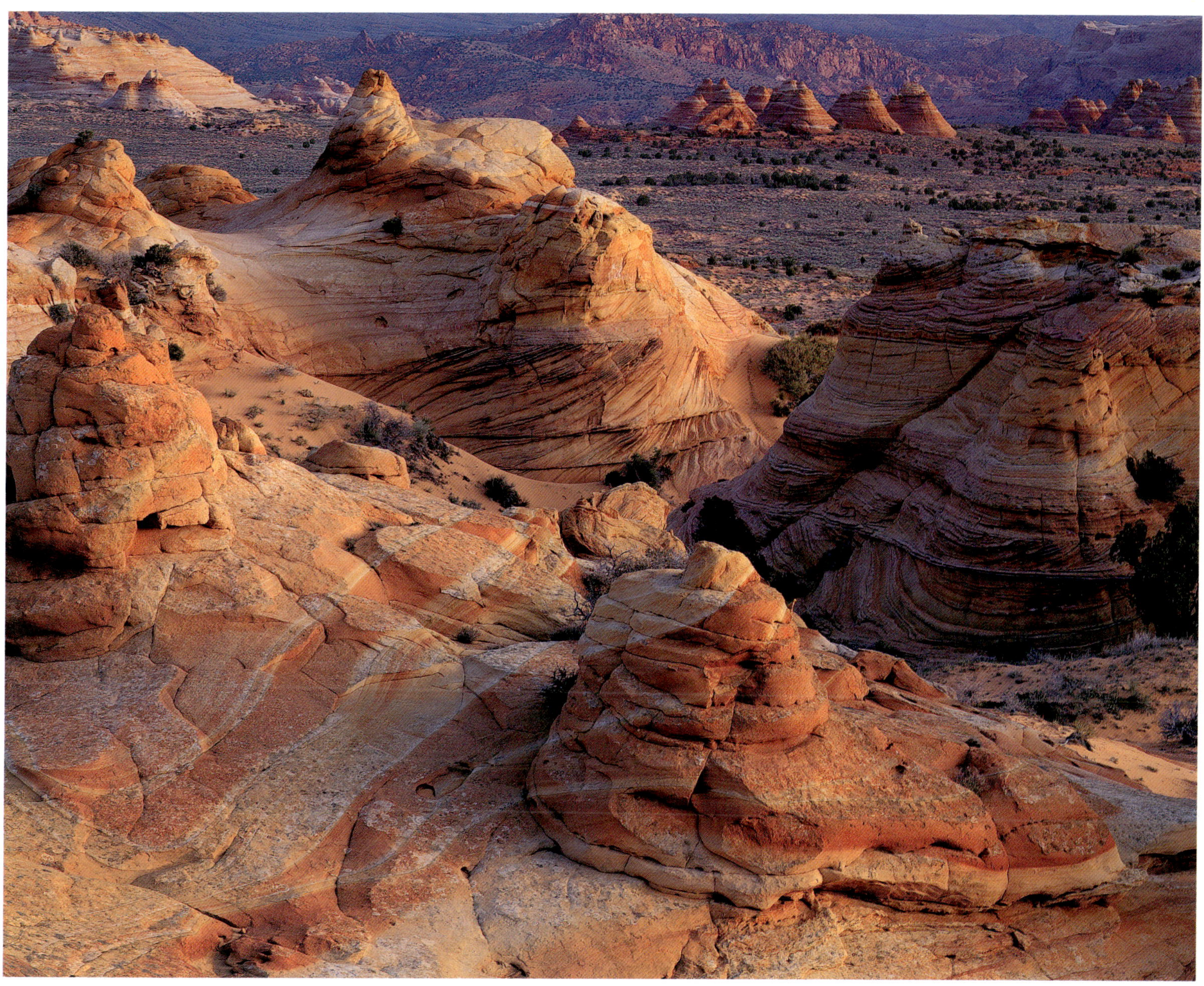

(above)

Eroded petrified sandstone dune formations at sunrise.

(Paria Canyon, Vermilion Cliffs Wilderness, BLM, Arizona)

(following page)

Ponderosa pinecones (Pinus ponderosa) *and driftwood amid canyon reflections.*

(Death Hollow Wilderness Study Area, Escalante Resource Area, BLM, Utah)

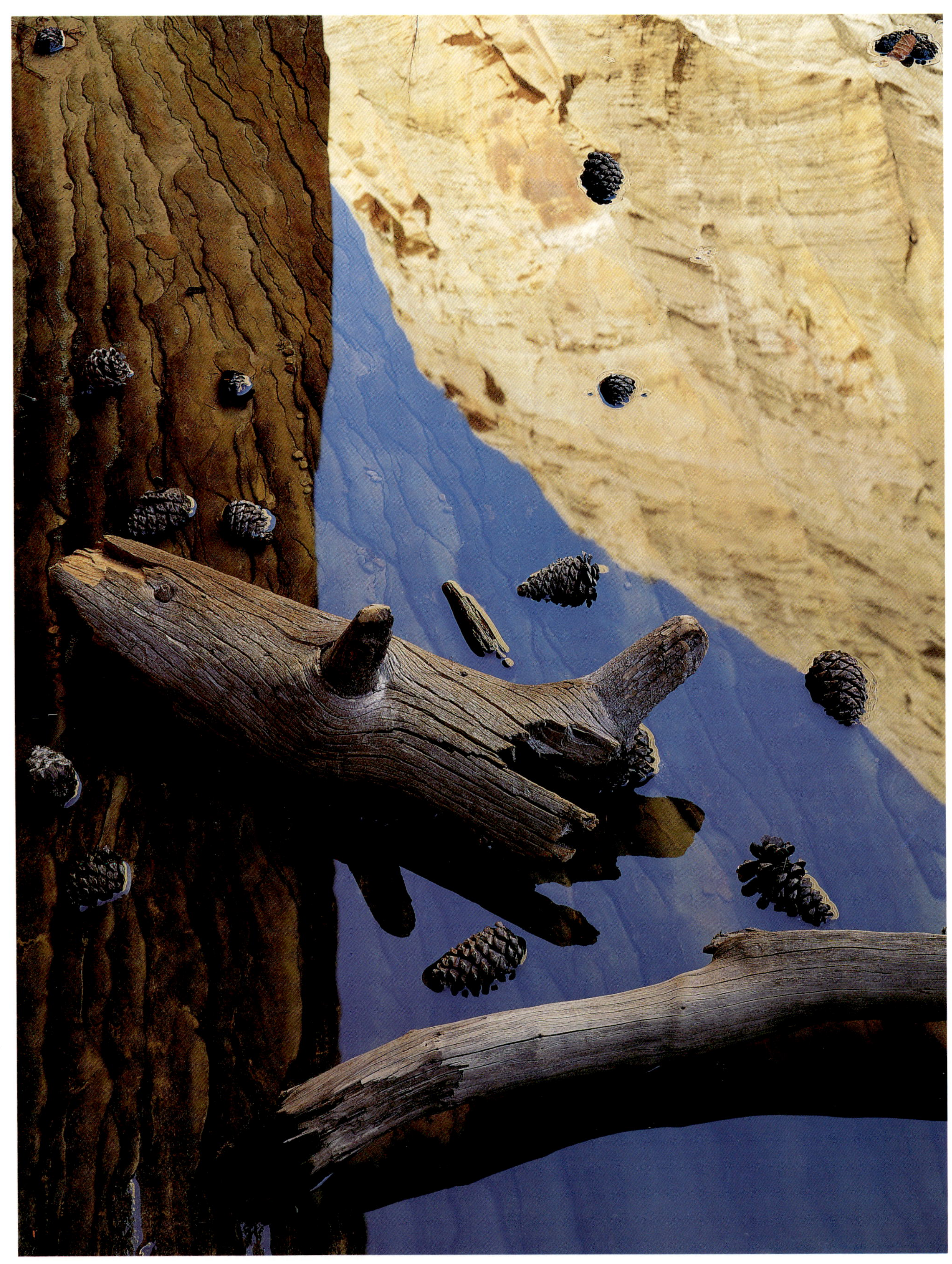

Contents

Foreword

In Charles Bowden's essay for this book, he writes, "You are going to a place that breaks your dreams, you are going into 'the heart of stone.' A hundred years after you die, men will walk on the moon, but few will walk in the country you plan to settle. Your failure will in time become a national treasure because it will remain silent and free and wild."

Bowden's words and the photographs of Jack Dykinga make it easy to understand why less than one hundred years ago, the canyons of the Escalante were virtually unknown to the outside world. The journeys they bring to life of the many doomed attempts of explorers and families hell bent on taming this wild land are as dramatic and extraordinary as the natural wonder of the Escalante itself.

Wallace Stegner once referred to the silence of this place as "the sort of silence in which you can hear the swish of falling stars." I find myself thinking about this often these days as the din of political rhetoric grows harsher, meaner, louder. I think about it as the sheer chaos of our cities, and often of our lives, force a fierce human struggle to find balance, peace, and the silence so essential to sorting it all out.

Uncorrupted natural places are a rarity in these modern times. Yet we hear so often that it's time we do something with the relatively few natural treasures left, that they have remained untouched for long enough.

What is it about these times that elicits such a chest-beating affront to anything pure, wild, and free? What is it about these times that manifests itself in such disregard for things greater than the sum of ourselves; intangible things that feed our souls, inspire our dreams, and nurture our hopes?

Must we "do something" with everything? I, for one, think not. What is pictured and described in this book only furthers my resolve on this. A battle to protect the kind of silence characterized by Wallace Stegner lies in the wake of the current political storm. Each of us who believes we are the better as a society for having access to that silence, must pierce another kind of silence—the silence of apathy—and let our voices be heard. Otherwise, the noise, the imbalance, and the chaos will be everywhere for the rest of time.

As Charles Bowden says, the failure to tame this land, will become a national treasure. A wonderful outcome indeed.

Robert Redford

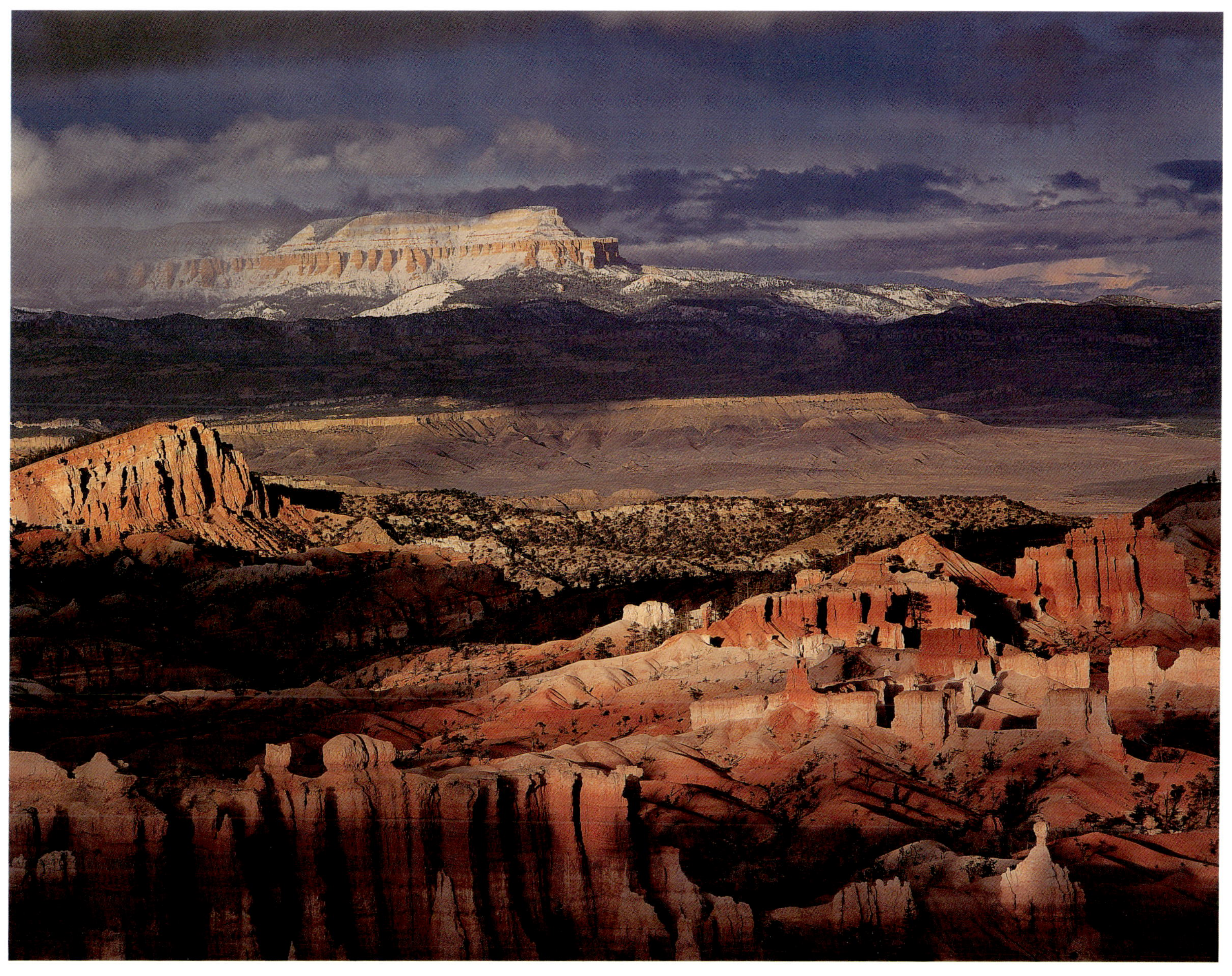

The spires of Bryce Canyon at sunset: new snow covers the Escalante Mountains in the background. (Bryce Canyon National Park, Utah)

Neon Canyon with circular spillways cut through the Navajo Sandstone. (Glen Canyon National Recreation Area, Escalante Canyon Complex, Utah)

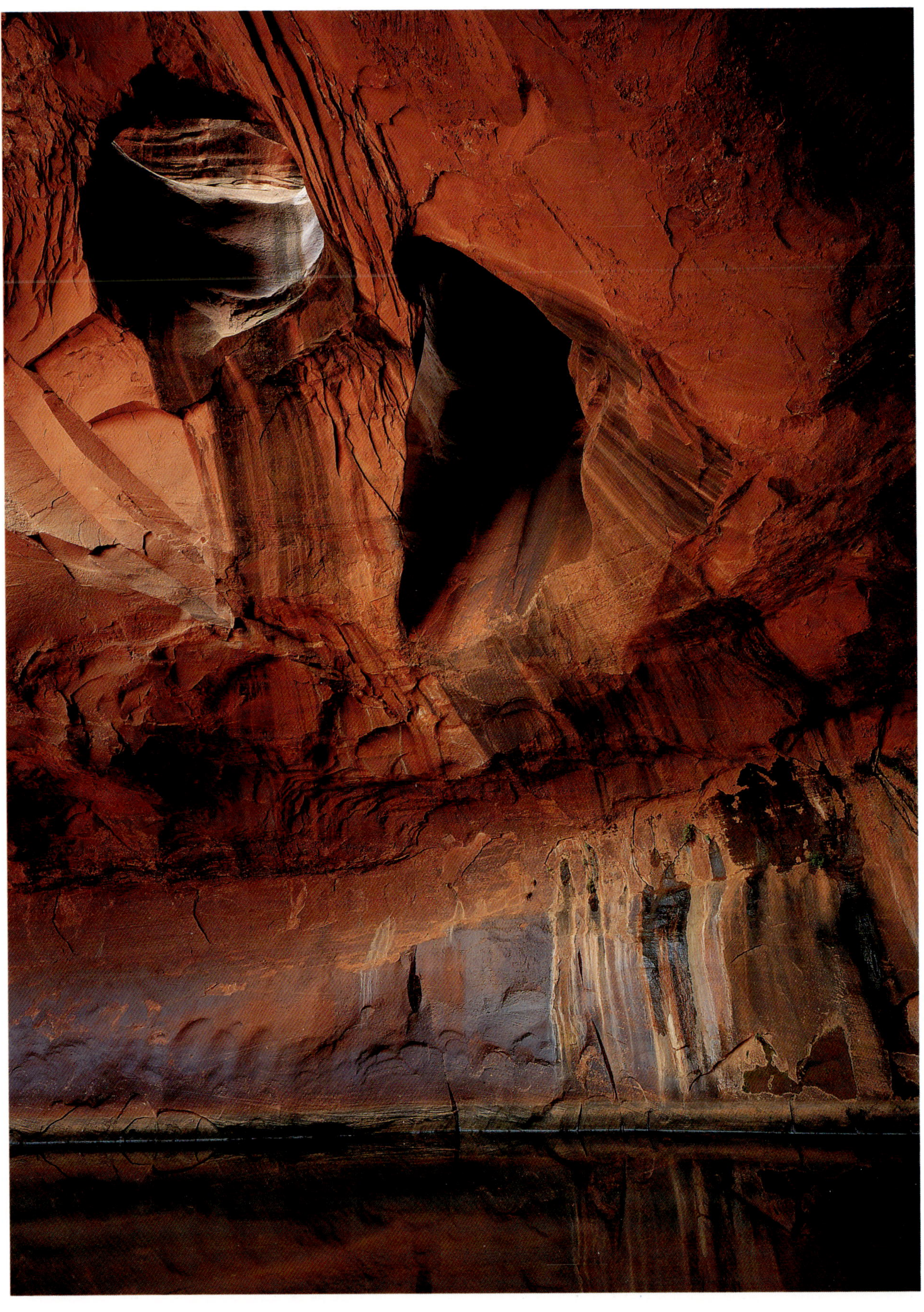

THEY WERE DETERMINED to be on time. This was not an easy goal to achieve because . . . they did not know exactly where they were going. Nor did they know who was going. Or when they were leaving. But, still, time was of the essence. They were a moving people, these Saints. During the 1830s, they had fled Kirtland, Ohio, for Missouri, the Central Stake that would anchor their faith and beckon them to return until the end of time. They had been driven out of Missouri and trudged on to Nauvoo, Illinois, in 1839. When their leader, the Prophet Joseph Smith, was murdered, they walked across the Great Plains and descended upon the Salt Lake Valley in the years 1846 and 1847. But the movement never really stopped. Their small farming communities multiplied and spread year after year and soon they had seized all the good ground in Utah, probed toward the Pacific Ocean with colonies called Las Vegas, Nevada, or San Bernardino, California.

Old Dreams

They repeatedly tested the stone paw of the big emptiness southeast of Salt Lake City. They tried to put down roots in the 1850s at a place called Moab, but could not make their outpost last. They looked into the country of the Paria River between 1865 and 1875 and found it wanting. In 1877, they were poised for yet another effort, but the death of their leader Brigham Young made them pause. But now they were ready and they were determined and they were insistent about being on time.

They prepared in this manner: it is December 28 and 29, 1878 and there is a meeting at the Parowan Meeting House and names are called. When you hear your name, you are to sell your house, sell your land, if single get a wife, and then load your wagon, drive your livestock, and go down that road to the place no one has yet determined. None of this will bother you because you are trained in founding towns and settling land and because you believe you are carrying out the plans of God Almighty. And so the names are called and the responses come in. You do not have to go, you can ask for and receive an exemption. But you are unlikely to shirk this task. Instead, if you are Henry Lunt, you will notice that the projected route trends east and you will say "that the march of the Saints today was toward the center stake of Zion [Missouri]. . . ." If you are Elder James Davis you will reveal that you have been "warned by a dream" that you would be "required to go & live in the Arizona Country. . . ." Or if you are Thomas Bladen you would simply say that you thought you could live "where anyone else could."[3]

This last statement is hardly a boast. The Saints are

the model of western settlers, the only arid-adapted people of European descent probing the region at that time. John Wesley Powell, when he finally delivers his *Report On the Arid Lands* to Congress in 1878, stresses the sensible nature of the settlements issuing forth from Utah. In a West of looting, drunkenness, land rape, and ruin, the Saints stand alone, with their irrigated agriculture carefully matched to the flows of the parched ground's rivers and streams.

They are not rovers or trappers or wild men or explorers. They are families pushing their nineteenth-century homes down the road. When they move they take their entire universe with them and their faith is so complete that they have few, if any, doubts. The wagon has a young woman in it and young children. Tucked away are the plow and seeds for the planting that will soon come—wherever they arrive they will start laying out fields the very next morning. Water barrels are strapped to the sides; ducks, chickens, and rabbits peer out from cages. There are needles and threads, tubs, grinding mills, soap, powder, bullets, and musical instruments. Nothing is left to chance. You cannot move down that road until the Bishop has checked your outfit. And behind this caravan dust rises from an immense herd of livestock.

You will refer to the expedition as perhaps the San Juan Mission, though there is still talk that your real destination may lie down in that fabled Arizona country. But you will soon be known by another name: the people who went through the Hole In The Rock. The trip is to take but six weeks, a lark. For months, this expedition has been the excitement in all the villages of Zion. The Church has various reasons for this migration. The country being settled is full of dangerous Native Americans (one winter 1,200 head of stock are rumored to have been stolen by them) and far more dangerous fellow citizens called outlaws. It is a void, full of what the faithful see as "bank robbers, horse thieves, cattle rustlers, jail breakers, train robbers . . . terrorizing and plundering inland settlements."[4] Also, there is a land rush going on and this rush is called the settling of the West. If the expedition does not take the ground, others will and they will likely be unbelievers.

There is an imperialism in the air, a strain so virulent that no one is aware of its existence. It is an unquestioned thing, which states that land is for the taking, that the present owners, Native Americans, have a feeble claim to this land and are not really using it anyway. Hardly anyone ever uses this word imperialism. They prefer to say civilizing. But the strain of imperialism is there and it can be carried in a wagon with the plow, the ducks, the chickens, and the rabbits. It is a matter of deep faith, almost a meta-religion, and it is seldom said out loud or questioned. It is not something that belongs to a particular class of American society or a particular religion in American society. Nor is it a quirk of certain regions. It simply is.

But what of the country into which you are about to

disappear? It is going to break your heart. There is barely a name for it in 1878, sometimes simply the San Juan Country and sometimes it is captured with a few sad words such as "it is certainly the worst country I ever saw. . . ."[5] Your journey with your young wife and your children and your livestock and your faith will not take the estimated six weeks. True, you will travel only two hundred and sixty miles, but you will be on the trail for six months. You will not be lost in some foreign country or in some dark continent. You will be at the heart of an exploding industrial nation, the United States of America. But you will disappear for months and months and you might as well be on the far side of the moon. You will have brought your imperialism, your absolute confidence in your ability to conquer land and turn it to account, to the one place in the republic that will forever repudiate you.

The last chain of mountains ever named in the continental United States, the Henry Mountains, escaped the attention of everyone until a few years before the start of your expedition. You are going to a place that breaks your dream; you are going into the heart of stone. A hundred years after you die, men will walk on the moon but few will walk in this country you plan to settle. Your failure will in time become a national treasure because it will remain silent and free and wild. No one will ever know exactly what to do with it. And then the day will slowly dawn when increasing numbers of people realize that nothing, nothing at all, is precisely what should be done with it.

The native people are called Paiutes, and they have a lesson to teach everyone. They refuse to live in this country, this heart of stone. At the jumping-off point live the Tuh'Duvaw Duhtseng, or Barren Valley People. To the south, on Navajo Mountain, live the Awdu'so Nengwoonts'eng band. To the east, in the Henry Mountains, are the Untaw'Duheutseng band.[6] They are wanderers, extremely resourceful at surviving in a landscape of low rainfall, scant desert vegetation, and irregular climate. They are basically indestructible. And yet they avoid the place you are going to. But then everyone else does, too. Of course, the native people will be at best ignored, at worst put on a kind of dole or slaughtered. Certainly, the knowledge lodged inside their minds will be given little attention. How can things be discovered if they are already known by others? How can mountains, rivers, valleys, canyons, and peaks be named if they already have a name? This work can only be accomplished by people who have drugged themselves into a vast denial, an ignoring. Then an unconscious imperialism is possible, one so potent it can discover ground already worn with trails and studded with ancient campfires. This is part of the process of making old country into new country. Declaring it new makes it possible to possess it and deny the claims of all others. Cartography, that hard, exact science of Europe, is rooted in this alchemy. This tactic has never ended—now it operates through the mouths of scientists taking vegetation and rock and fragments of vanished people and by this act of naming making them once again the property of those who have come later, after some earlier flood of life. It is in our blood, then, now, tomorrow. Apparently, we cannot be cured or will not stay the course in the therapy centers. But of course the naming is not the real issue—it is a false triumph, like the fiction of discovery. Understanding is the bottom line, whether the person is a Paiute huddled in a grass shelter against the wind or an ecologist riding in a helicopter during an environmental survey. And in this struggle, the one toward understanding, words like triumph and conquest avail us not at all.

This area has a possible name, one that floats on the surface of its river. It is called the Escalante. But even the name is a trick. The man who left his name on this river and this country never really entered it. He peered into it and fled. That was long ago, a century or more before this expedition of the Saints, and now his journey is buried in Spanish archives. But his conclusion will come back to haunt you and all who come after you, generation after generation. The great boat expeditions of John Wesley Powell down the Colorado River missed even noting the mouth of the Escalante, and when it was finally named in 1872 it became the last river added to the map of the continental United States.

That is the lesson of this book. The Escalante matters because it has never mattered. This is the ground that has repudiated our history. We scratch at the rocks, leaving graffiti to mark our passing, but we have no real consequence here. The miracle of the desert never occurs here. The rise of cities does not happen here. The railroads stay away, the vast interstate highway system skirts this stone in alarm. Most of the wells come up dry; all the schemes collapse into ruin. And we are hopefully left with the grace of life itself.

But of course, this has not yet happened, it is simply something beginning to happen. Today, exploration parties still probe this layer cake of ancient stone for oil or natural gas. Large beds of low-grade coal are ripped from the ground to fuel the electrical dreams of the Southwest and the Pacific coast. We are almost addled in these quests. The Colorado Plateau, that huge cake of rock that incorporates the Escalante drainage and the Paria drainage and much more, has never really worked out for us. It is an affront to our very nature to have so much of our ground . . . so worthless. We are not by nature a complacent people. We think, if we bide our time, something will turn up. We share the faith of our ancestors, though we now speak more softly of these matters. Our faith has become something of an embarrassment to us, but still it lingers deep within us.

The huge counties of the Colorado Plateau in Utah today host 28,000 people in a place larger than Massachusetts, New Hampshire, and Vermont combined and this handful of folks live in but a few towns and hamlets. We still keep the faith, we still keep looking for that something, but in our hearts we all truly know the failure and this fact either brings a faint smile to our faces or the dark clouds of rage.

We have become crafty at disguising it, but its primary tenets still ring true to us: the earth exists for us, and human ingenuity will make this earth perform according to our desires. Now we have management plans—national forests, national monuments, recreation areas, official wilderness areas, national parks, wild and scenic rivers—but this is part of our disguise. We backpack, float rivers, ride mountain bikes, take survival training. The impulse is the same, things must exist for us or why do they exist at all? What if they exist simply to remind us of our insignificance? What if their function is to be functionless in our eyes? What if they are there to disdain us and teach us humility? Behind our place naming, peak bagging, rapids running, picture taking, back there where we seldom look and do not like what we see when we look, what if in that place we are irrelevant, and our busy nature adventures are but a way for us to deny our unimportance? I am alive, it is nearing the end of the twentieth century—a unit of time that would make the very rocks of the Escalante country laugh and shake with mirth if we bothered to tell them—and there is this scene we must consider. . . .

THE GROUP HAS HIKED in to get away from it all, deep into the stone where their sacred permits promise they will meet no one. Then a pack train arrives carrying a ton of gear, a fifty-gallon drum of water, camping supplies, food, and a couple of assistants for the photographer. He has a massive camera and he has crossed the nation to take a picture just like the picture he holds in his hand—one published by yet another photographer. As I asked: what if we are driven crazy by our own irrelevancy here; driven to a state that drugs cannot address and we cannot abide, so we assume a behavior to mask what we really feel and dread? Perhaps there is very little difference between ourselves and our ancestors, except the lengths we are willing to go to disguise the anguish we feel when we face this beautiful place that cares about us not at all. Cosseted in our Lycra and our polypropylene we carry on bravely, and thus can strangle the voices from the past warning us, whispering caution as we plan our second home, and decide on the gear ratios for our special bike. . . .

I am sprawled on the ground and it is very hot lying here on the sand. A few hundred yards away, the river purrs in a deep canyon, and the walls are tan and pink and brown. Huge stone arches cavort in Davis gulch off the main stream of the Escalante. The brown tongue of the river rolls lazily between the young cottonwoods and willows lining its banks. At my back, the Kaiparowits Plateau makes a fifty-mile wall of stone. Across the river, the Waterpocket Fold guards the east and northeast approaches. The silent river flowing below has come oozing down from the Aquarius Plateau and soon will vanish into the dead waters called Lake Powell. The morning light is very bright and there is no shade on this flat high above the river. I get up and amble between hummocks of sand and then scamper up the soft form of a rounded rock outcrop. Here I find the last gasp of evening primroses in small pockets of shade, the big white flowers defying the huge sweep of waterless desert. The wind has come up now and there is nothing to stop it. Behind me a few miles to the west is the route taken by the expedition in 1879. I can hear the barking of their dogs, the lowing of their cattle, the voices of the men and women and children, the creaking of the old wagon wheels fighting through the endless sand. Such moments are not much of a reach here and require very little imagination. None of the changes—the dams, faint scratches of roads, powerlines, jet aircraft slicing the sky thirty thousand feet overhead—have really changed the fundamental conditions. All the yesterdays are very close, because the ground refuses to acknowledge the passage of time, because the ground ridicules our puny efforts at conquest and power. So I stretch out there on the rock by the blooming primrose and listen to the distant voices as they march on toward failure. I can hear them singing, "Come, Come Ye Saints," and the song tugs at my heart. They are doing what Americans know how to do. They are doing what we all believe in and cherish; they are building, forming, taming, growing, making their mark. And they are doing it in the wrong place. This is not an easy thing to say when you are lying there by the blooming rose and hear the hope and warmth in those voices. Two hundred and fifty human beings, eighty wagons, hundreds of head of cattle and horses, constant singing, all heading into a heart of stone.

They only have to travel one hundred and thirty miles as the crow flies. But they are not crows. And this is not likely to be Zion.

Into the Heart of Stone

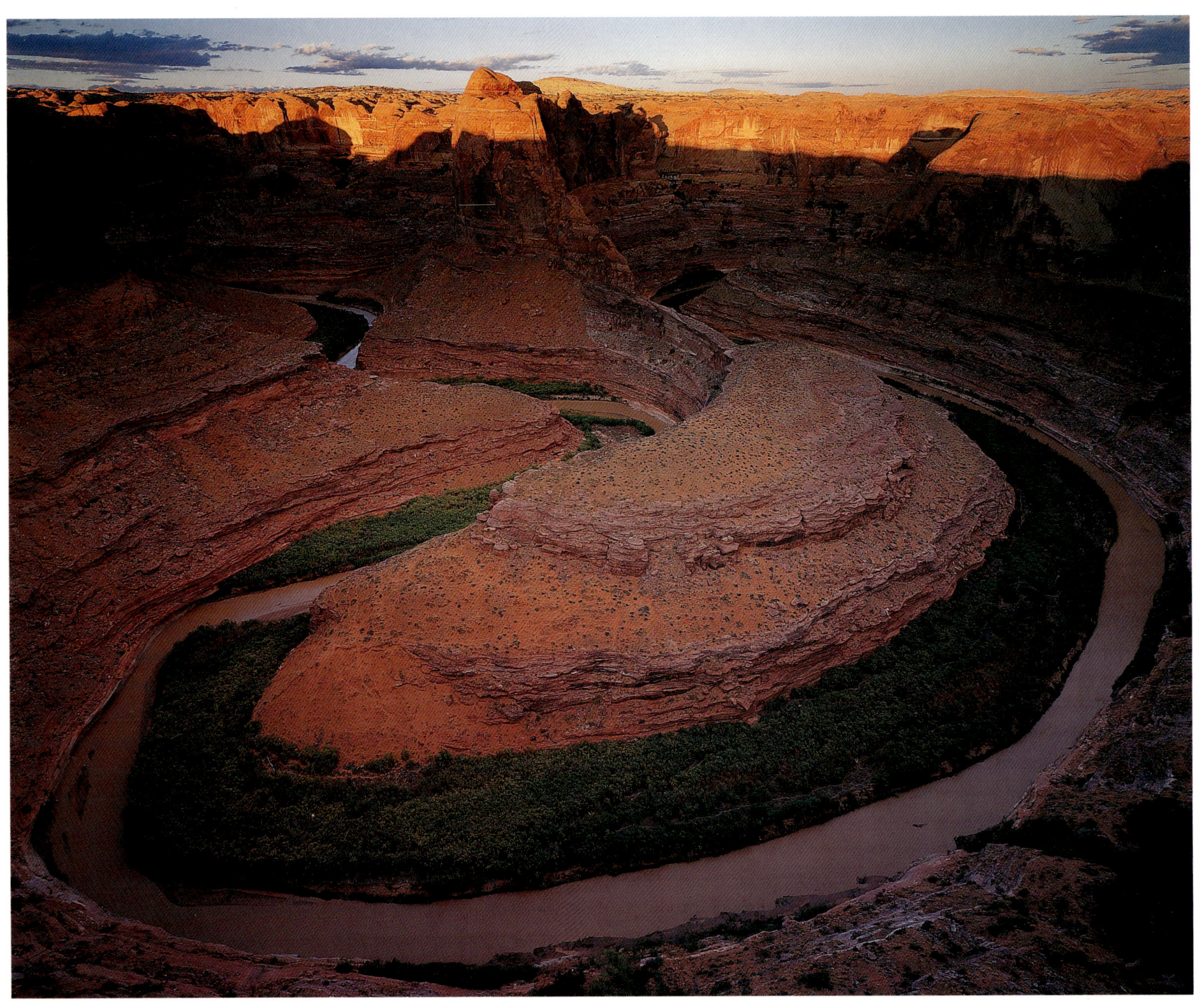

The Escalante River winds its way toward Lake Powell and the Colorado River. (Glen Canyon National Recreation Area, Utah)

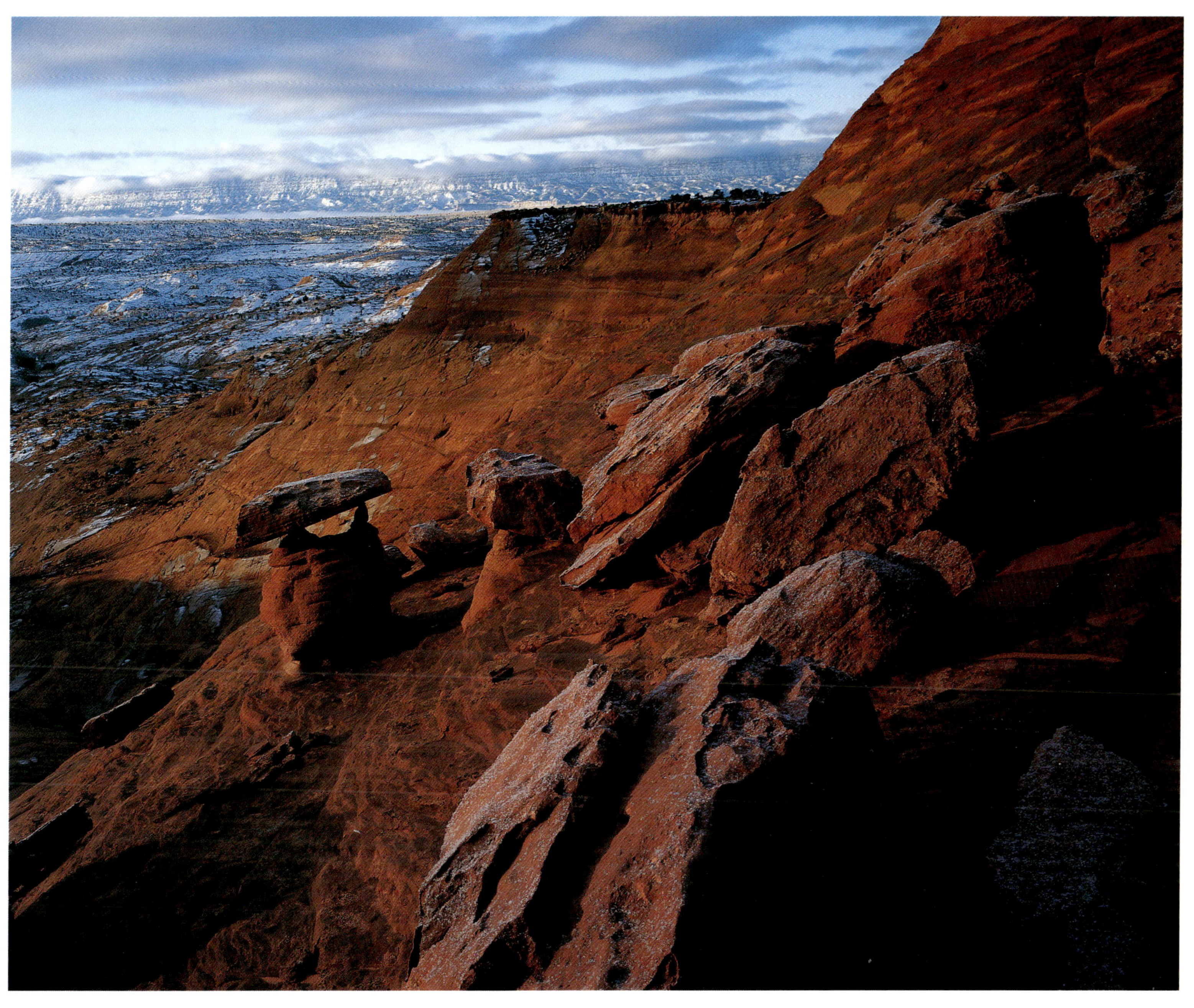

Near 25-Mile Canyon, rim rocks are dotted with ice and a snowstorm hangs over Straight Cliffs in the background. (Escalante Wilderness Study Area, BLM, Utah)

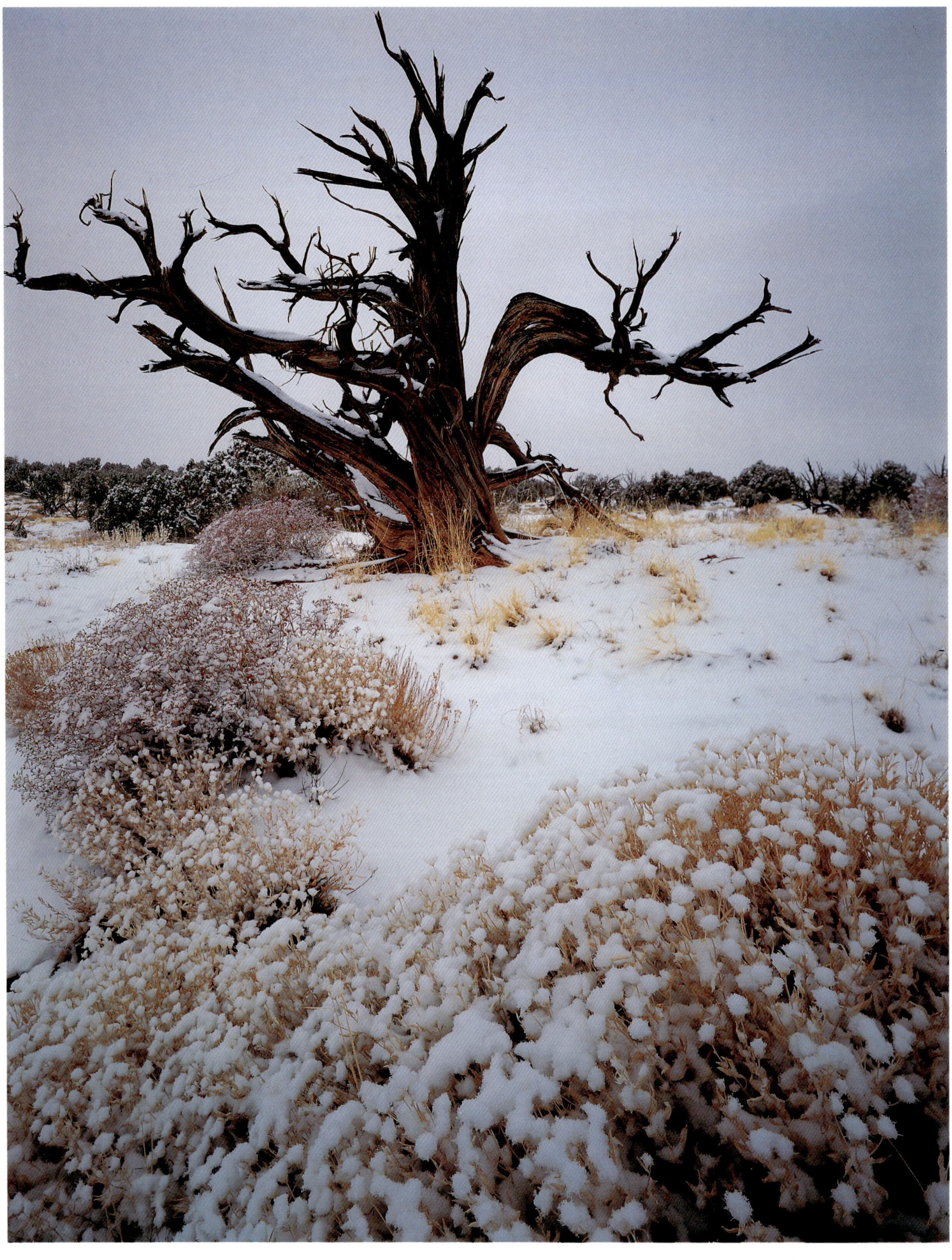

Utah juniper (Juniperus osteosperma) *skeleton with snow covered wild buckwheat and other plant species. (Escalante Resource Area, Wilderness Study Area, BLM, Utah)*

Skeletons of Utah juniper (Juniperus osteosperma) *stand out against clouds at sunset. (Escalante Resource Area, Egypt Bench, BLM, Utah)*

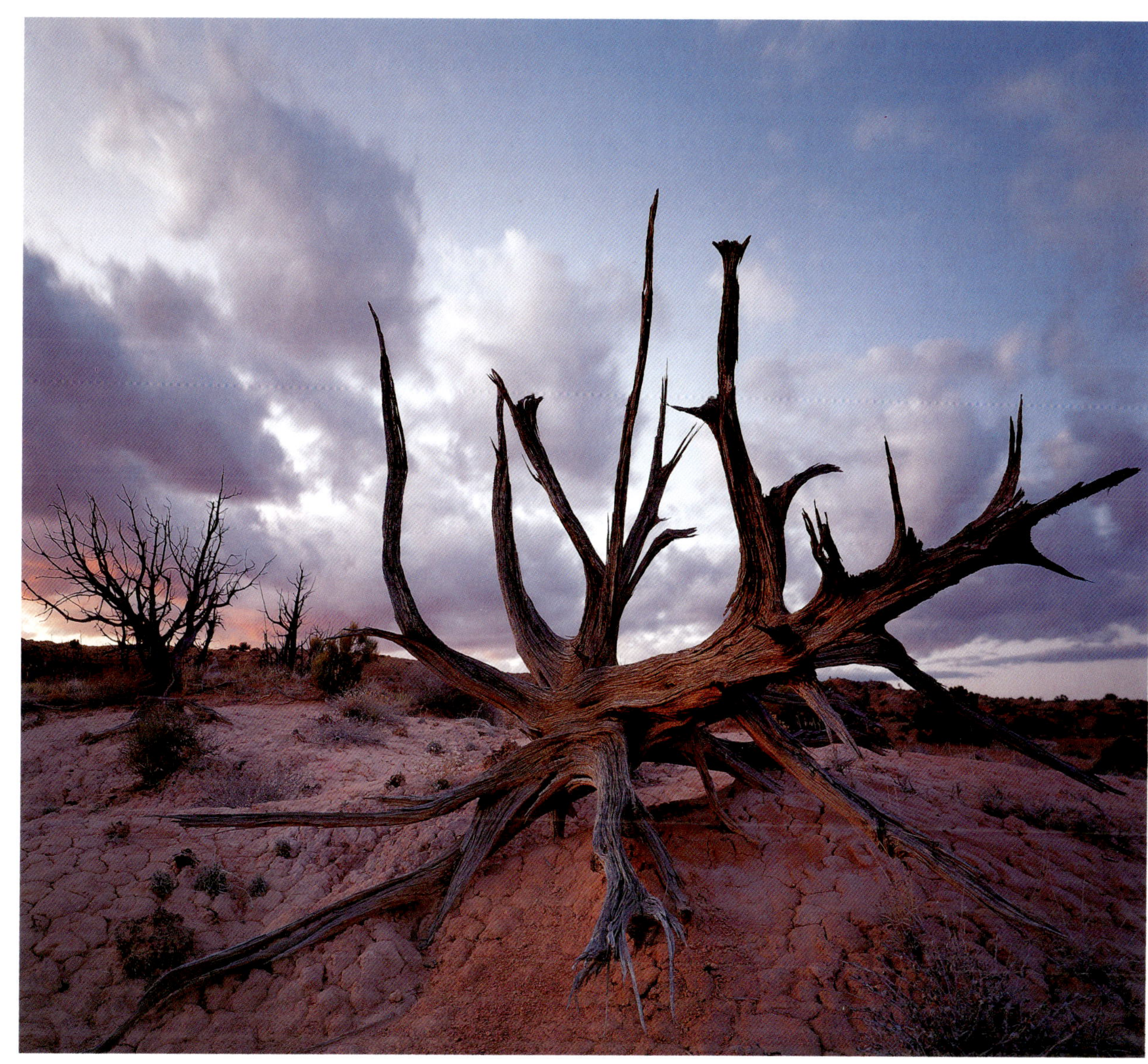

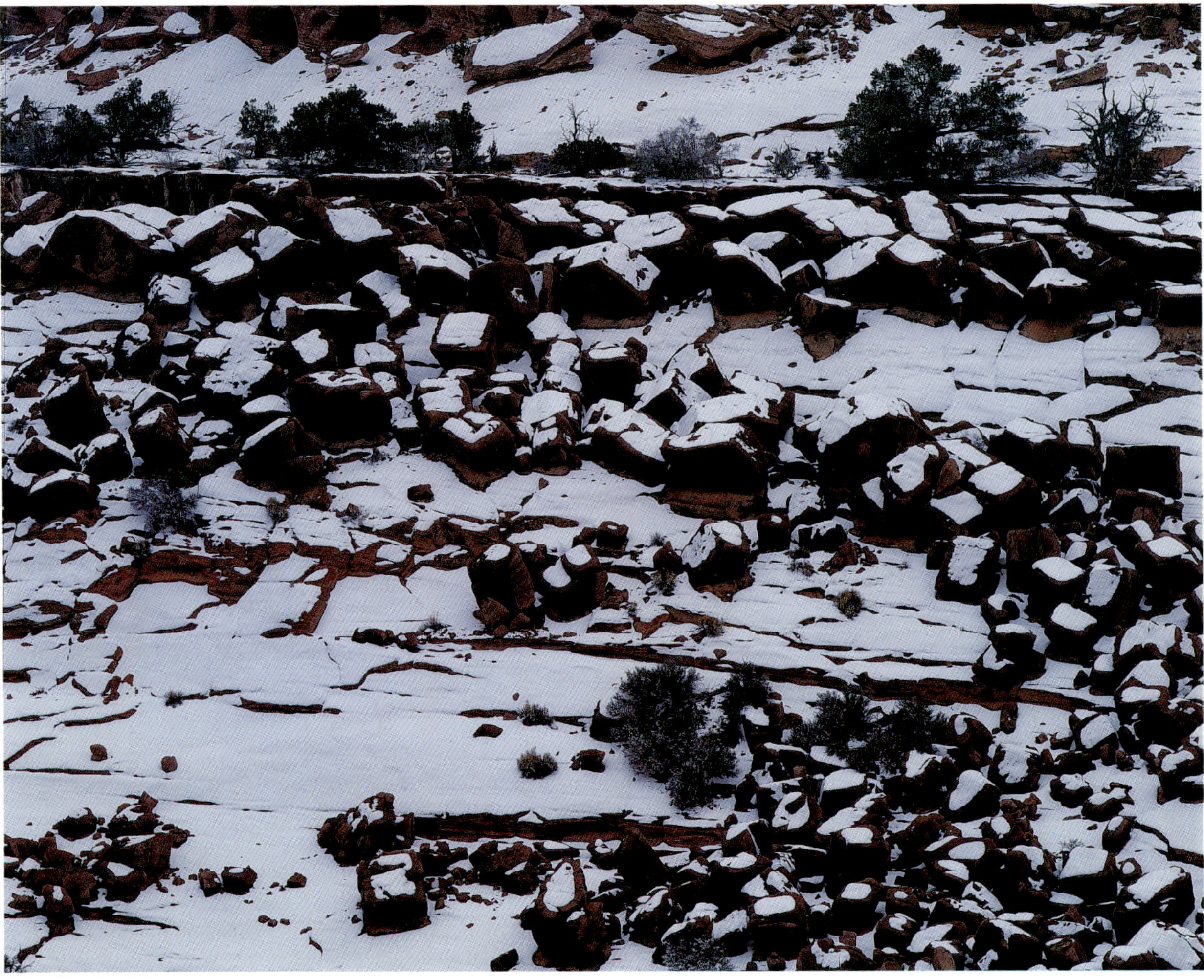

Crumbling blocks of rim rock covered with fresh snow create an abstract geometry. (Escalante Wilderness Study Area, Utah)

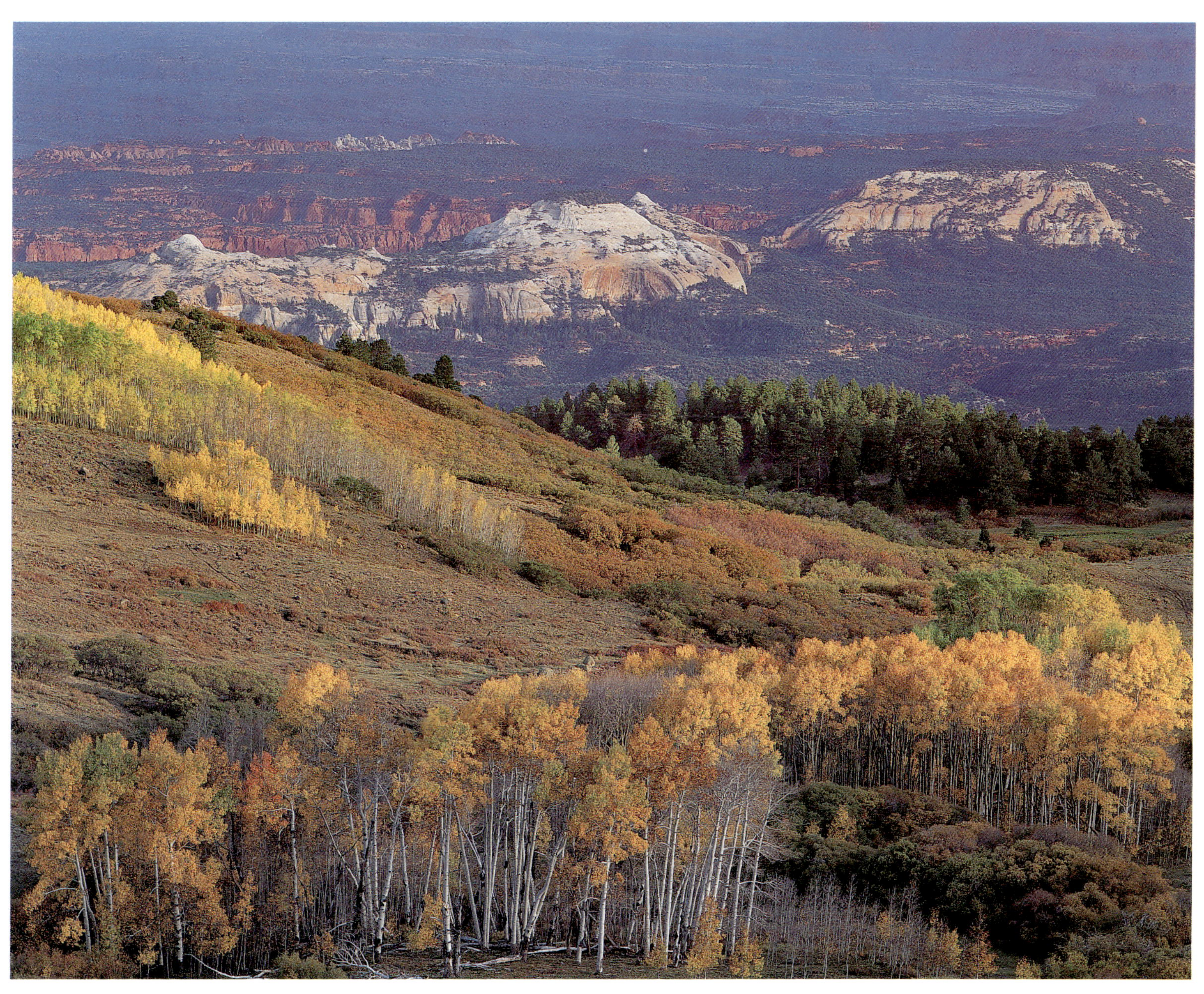

The Escalante Canyon complex in evening light. (Dixie National Forest, Boulder Mountain, Aquarius Plateau, Utah)

Metate Arch with sunset light on the Navajo sandstone formations. (Escalante Resource Area, Devil's Garden, BLM, Utah)

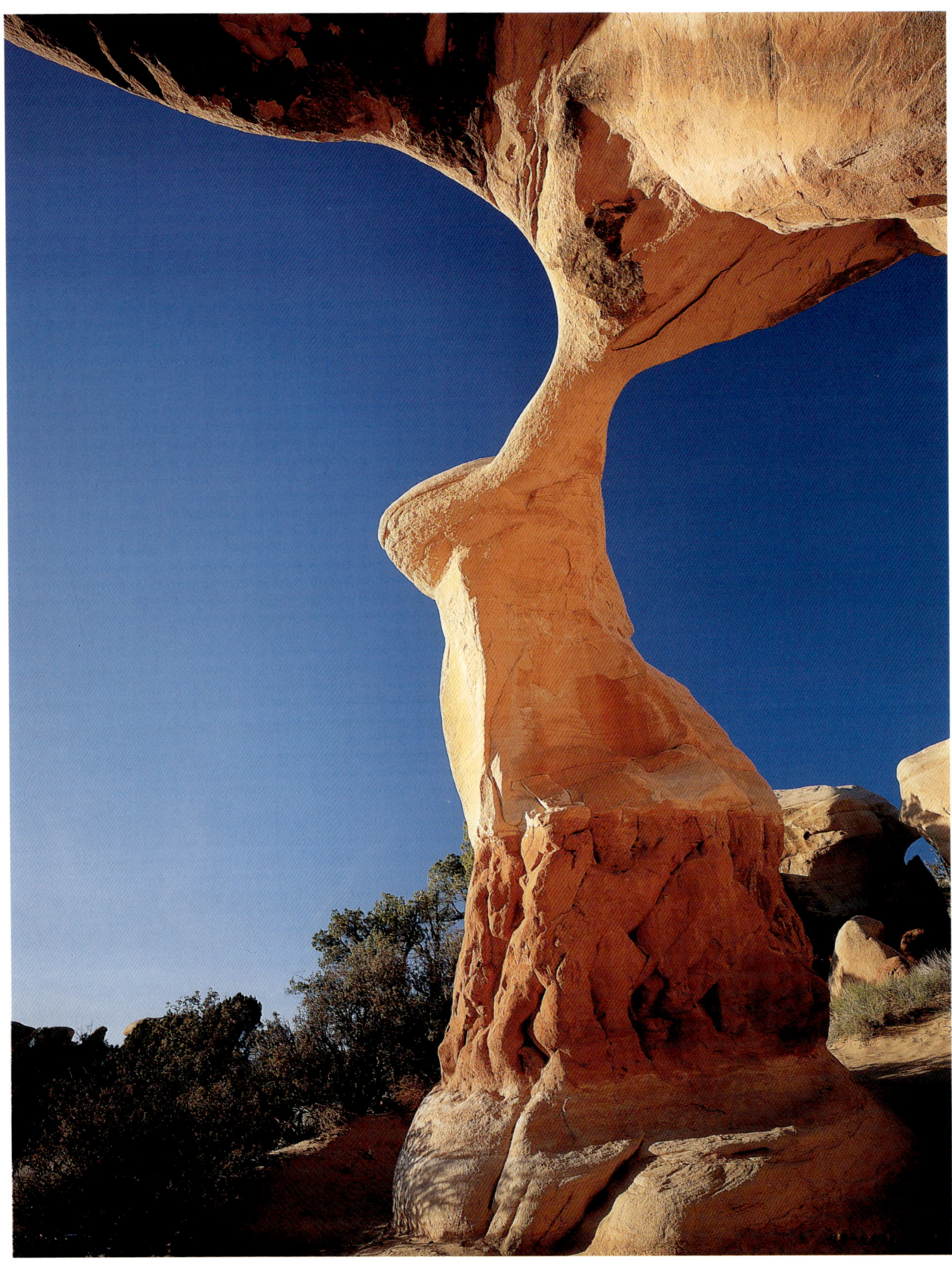

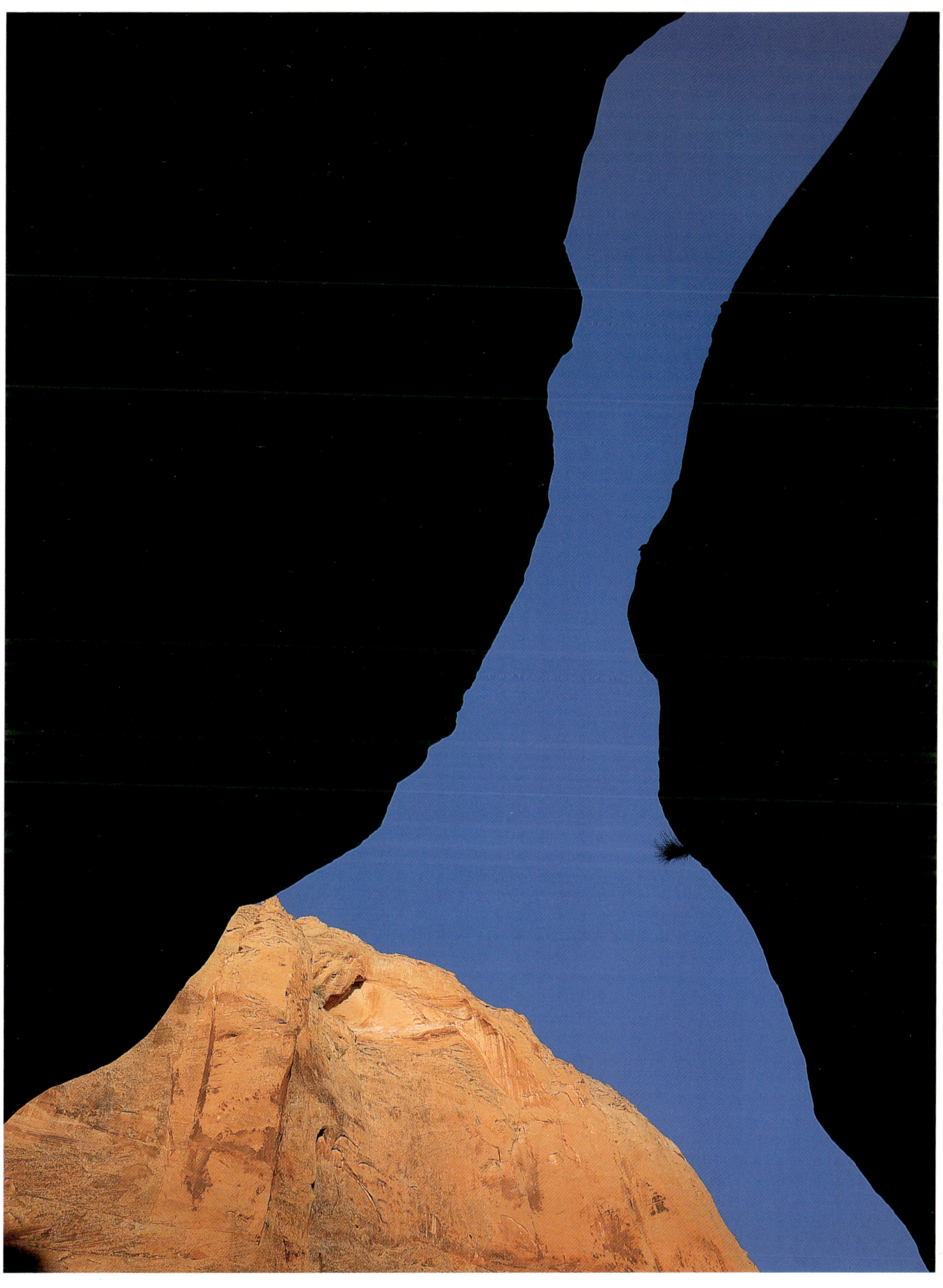

Canyon walls near the junction of 40-Mile and Willow canyons. (Escalante Canyon, Glen Canyon National Recreation Area, Utah)

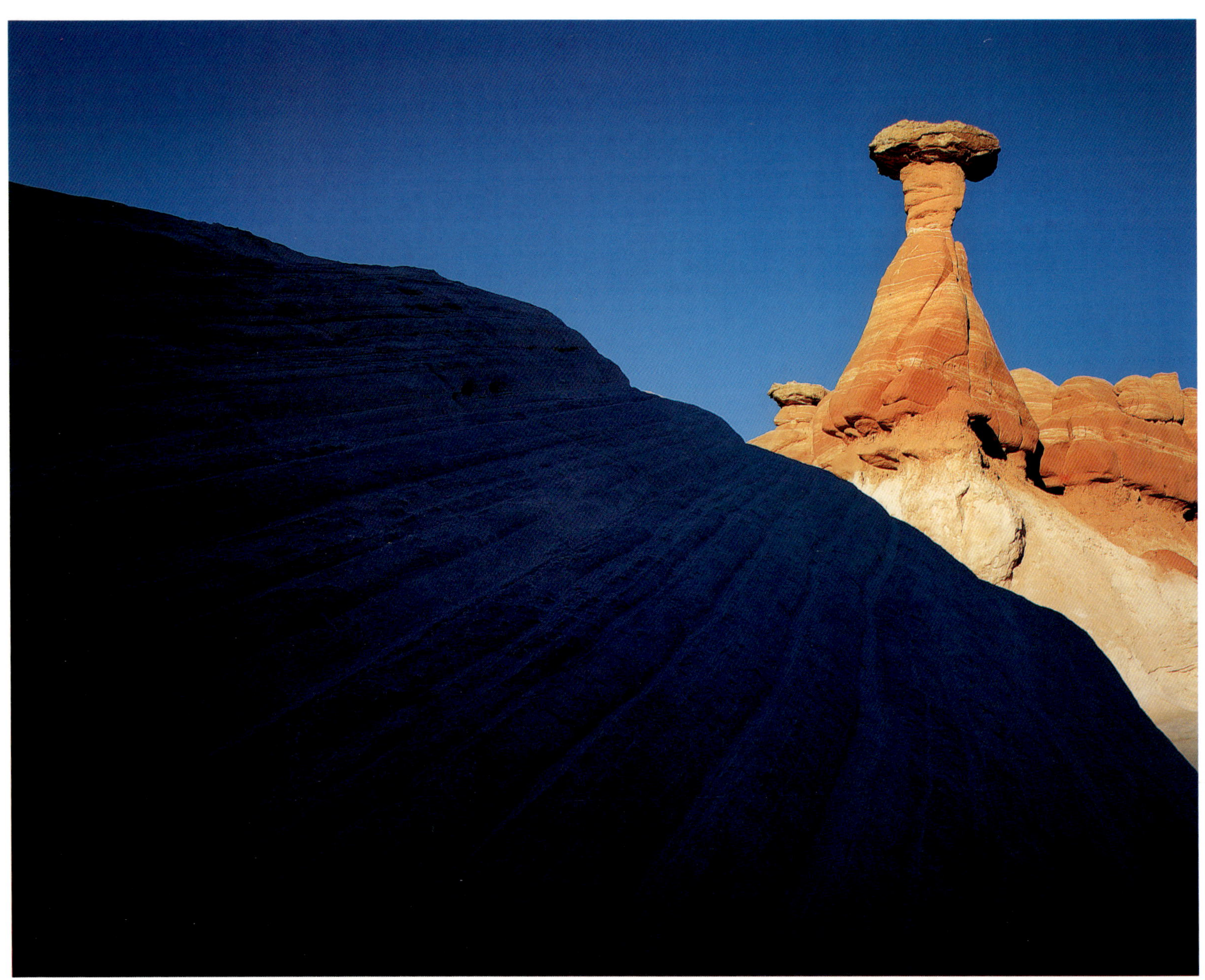

Contrasting sandstone with "Hoodoo" cap-rock formations in background. (Cockscomb Wilderness Study Area, BLM, Utah)

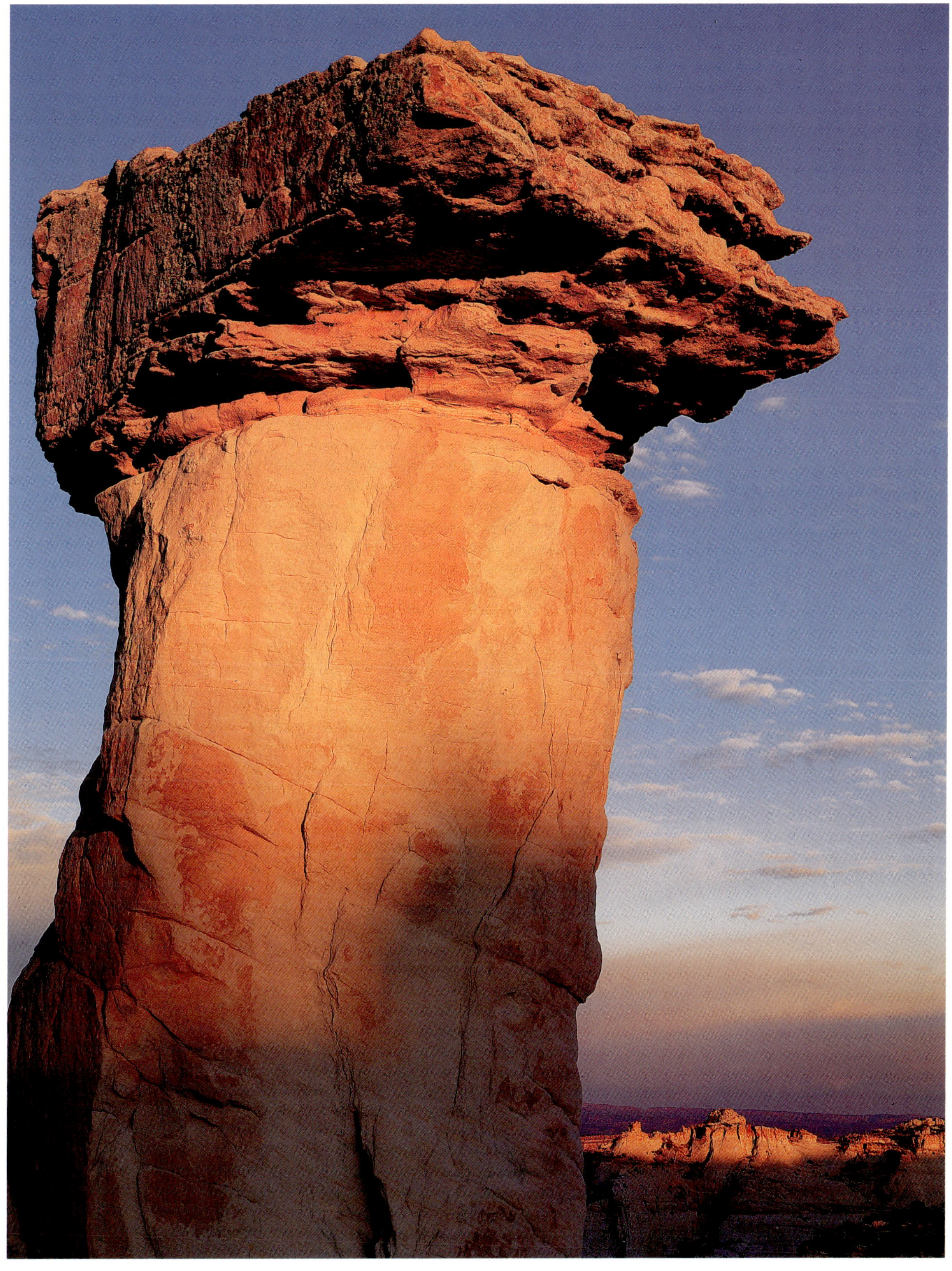

Navajo sandstone with Carmel formation cap-rocks at sunset. (Rim of Escalante drainage, Wilderness Study Area, Escalante Resource Area, BLM, Utah)

Sandpaper mule's ear (Wyethia scabra) *flowers in the Entrada sandstone formation. (Sooner Rocks, Escalante Resource Area, BLM, Utah)*

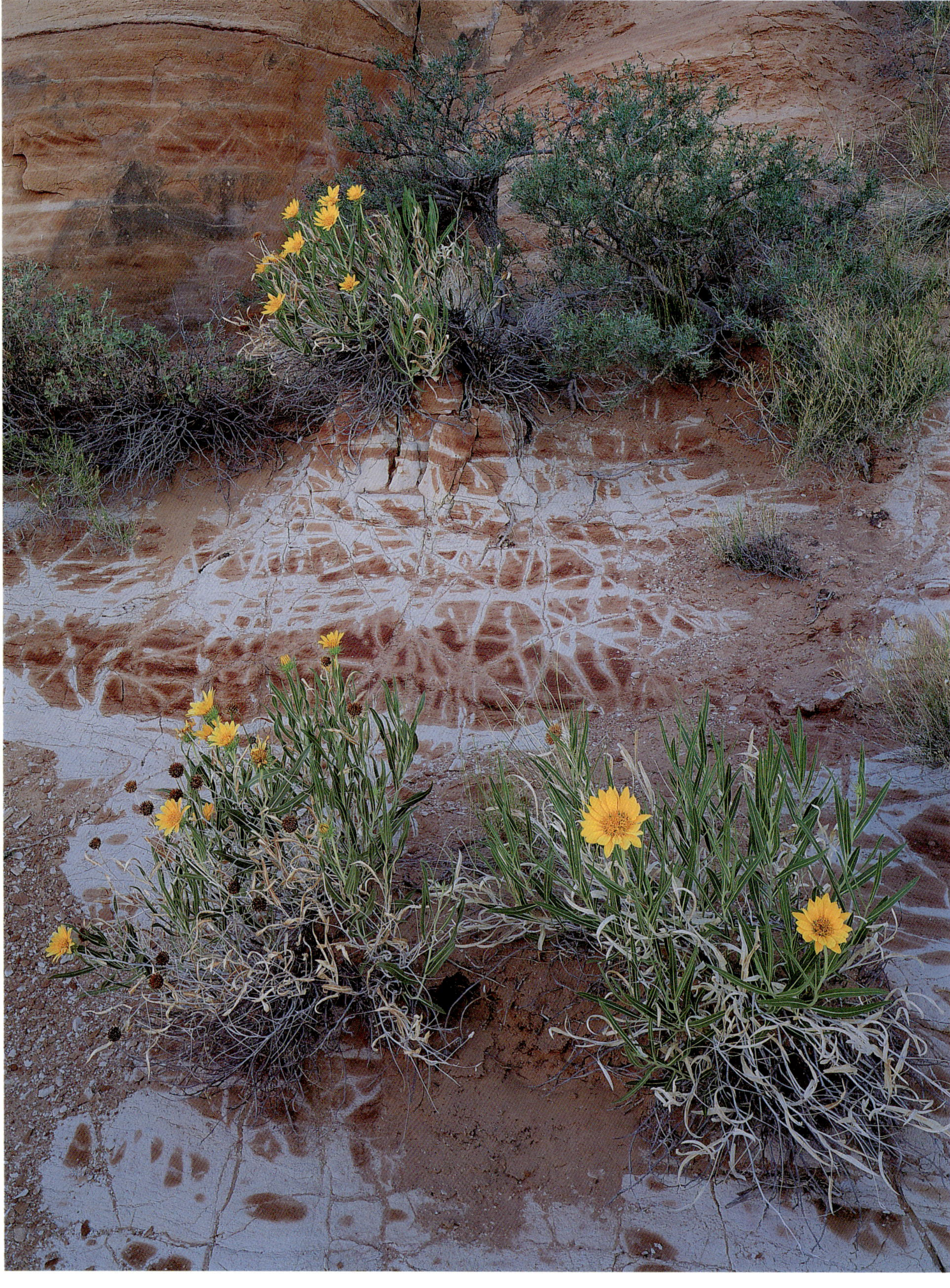

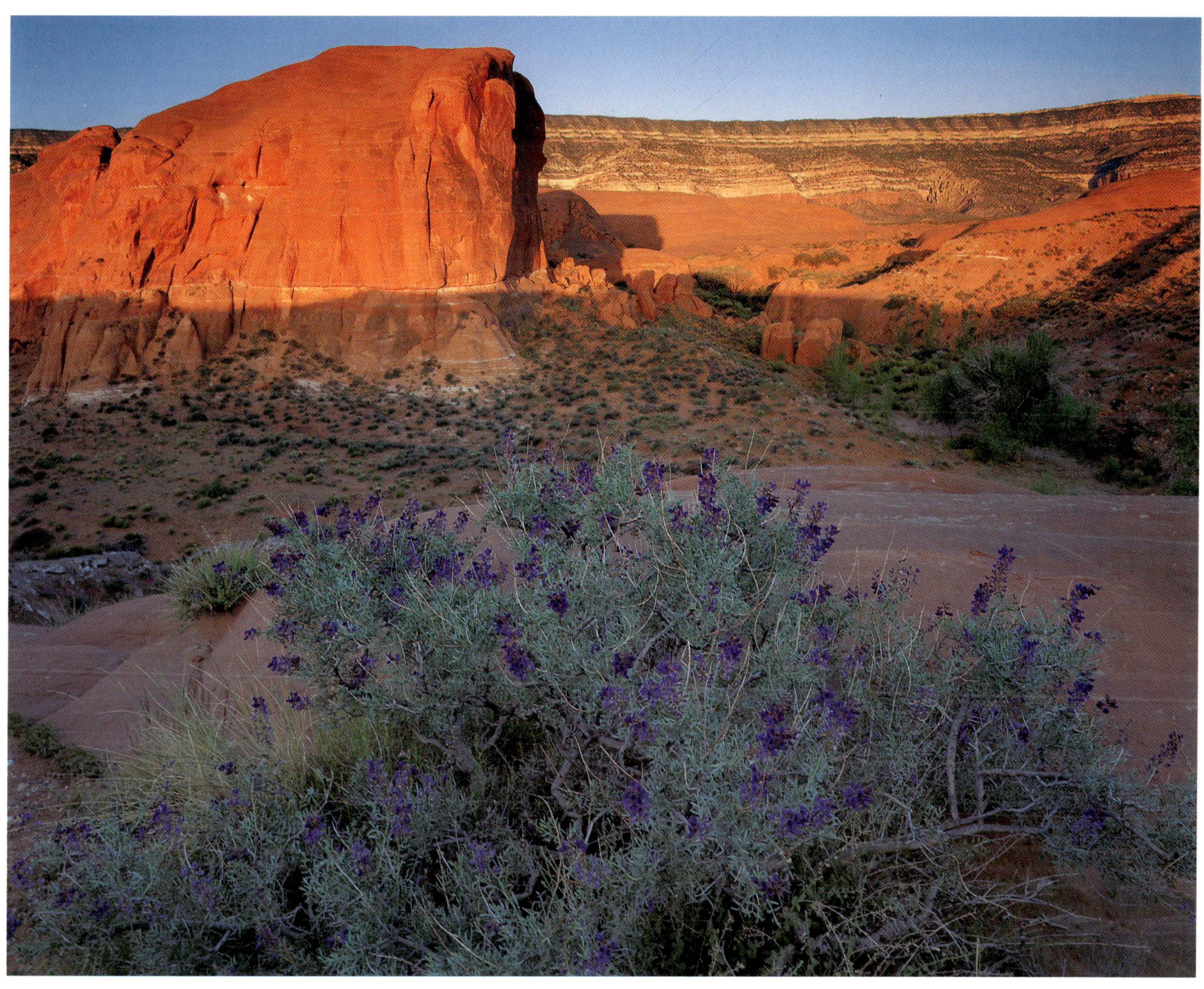

Sooner Rocks, an outcropping of Entrada sandstone at sunrise, with indigo bush (Psorothamnus fremontii) *in the foreground. (Escalante Resource Area, BLM, Utah)*

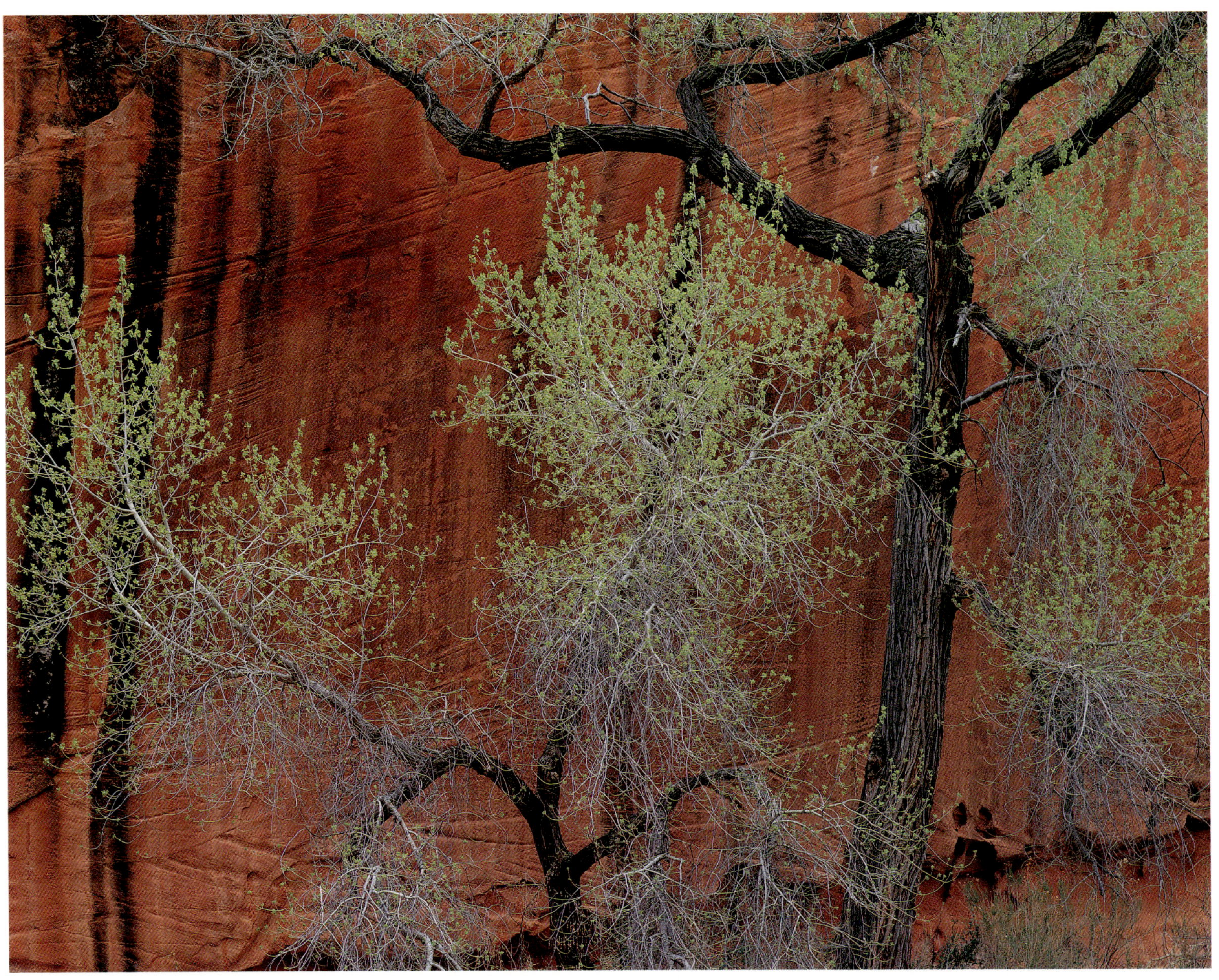

Spring foliage of a cottonwood tree (Populus fremontii) *seen against Navajo sandstone. (Escalante Canyon, Glen Canyon National Recreation Area, Utah)*

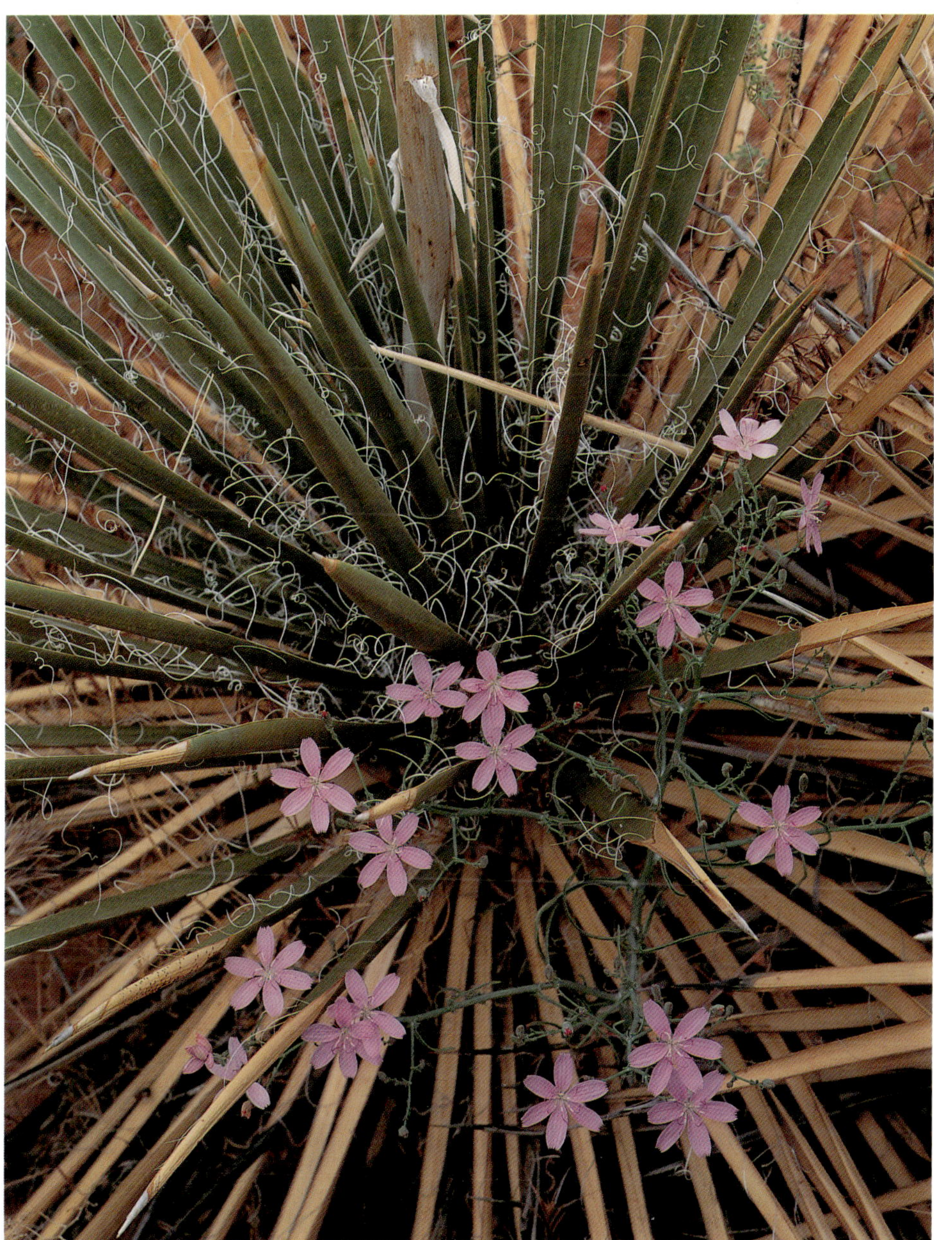

Banana yucca (Yucca baccata) *grows beside flowering wire lettuce (Stephanomeria sp). (Davis Gulch, Glen Canyon National Recreation Area, Escalante Canyon complex, Utah)*

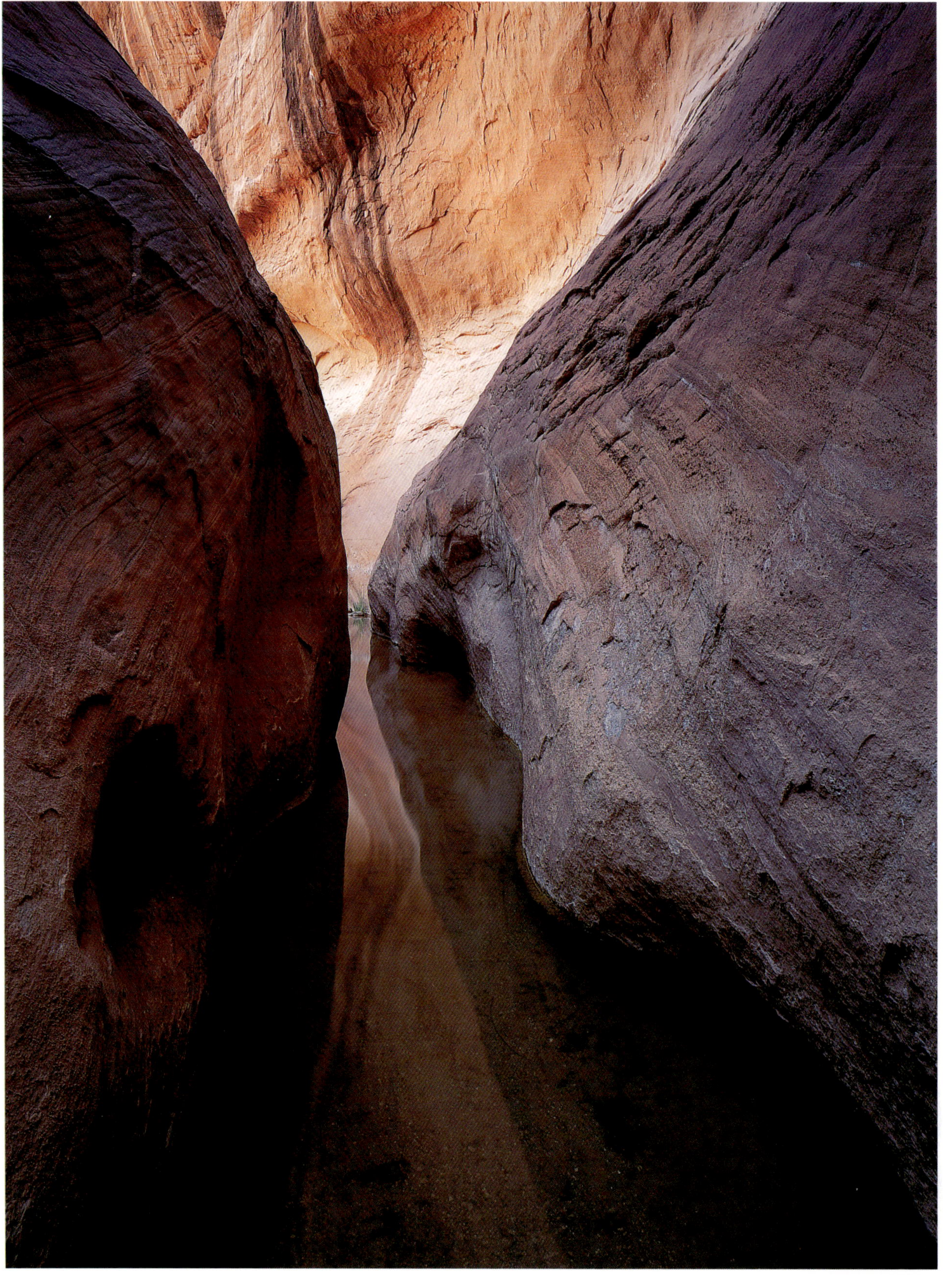

Reflections of the Navajo sandstone in the narrows of Willow Canyon. (Glen Canyon Recreation Area, Escalante Canyon complex, Utah)

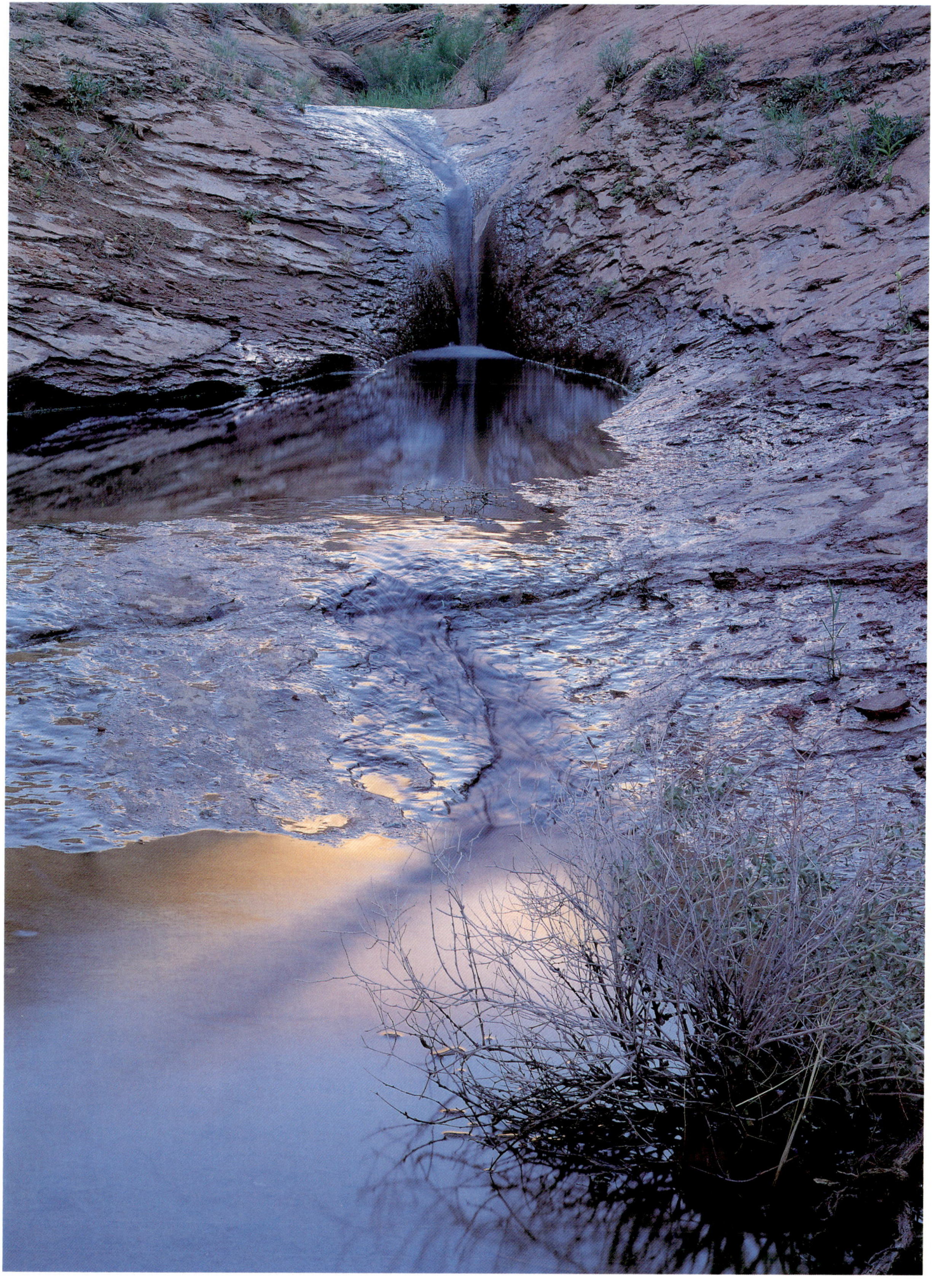

Uprooted wild buckwheat (Eriogonum corymbosum) *appears below a waterfall in Willow Canyon. (Glen Canyon Recreation Area, Escalante Canyon complex, Utah)*

Contrasting sandstone with "Hoodoo" cap-rock formations in the background. (Cockscomb Wilderness Study Area, BLM, Utah)

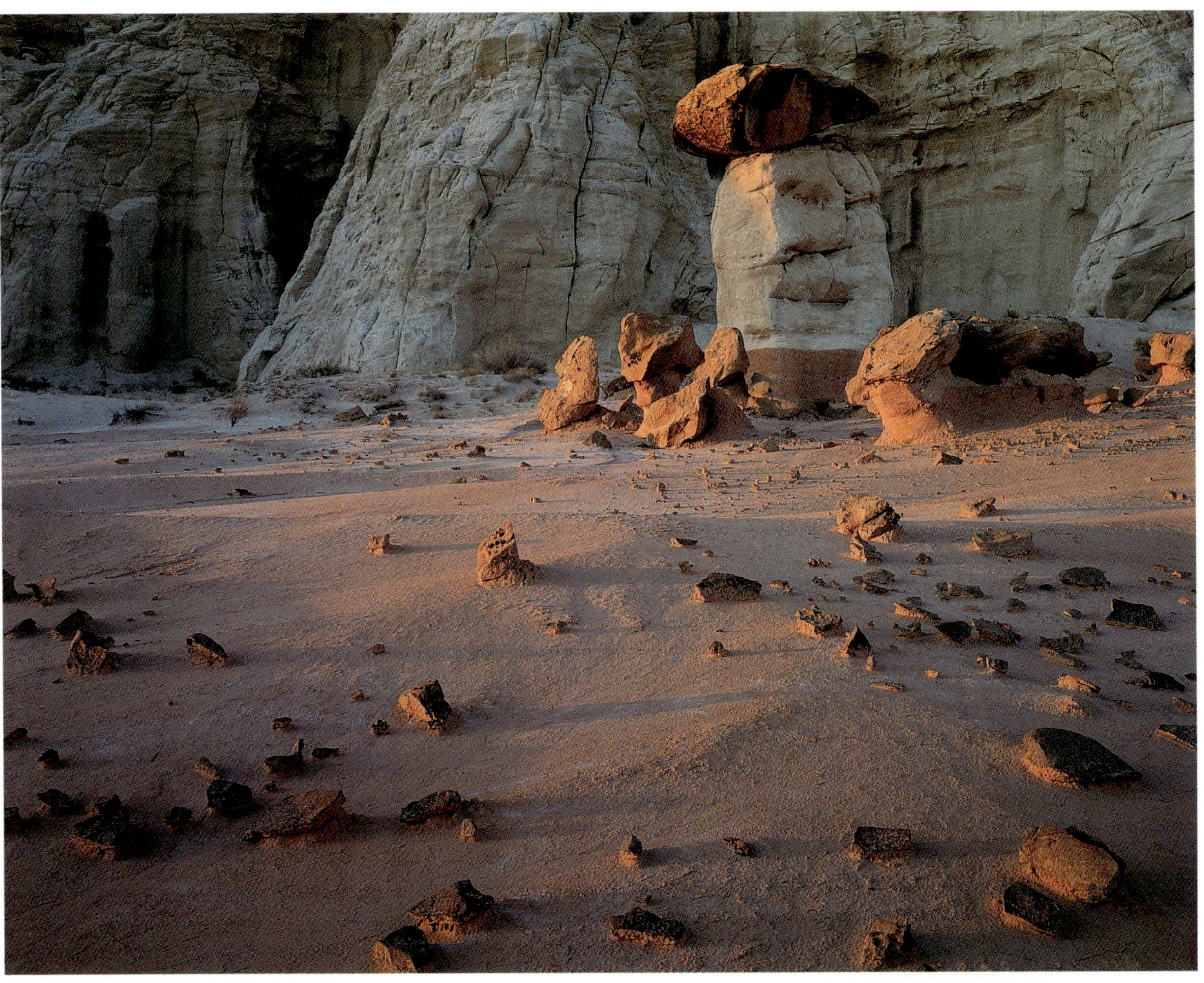

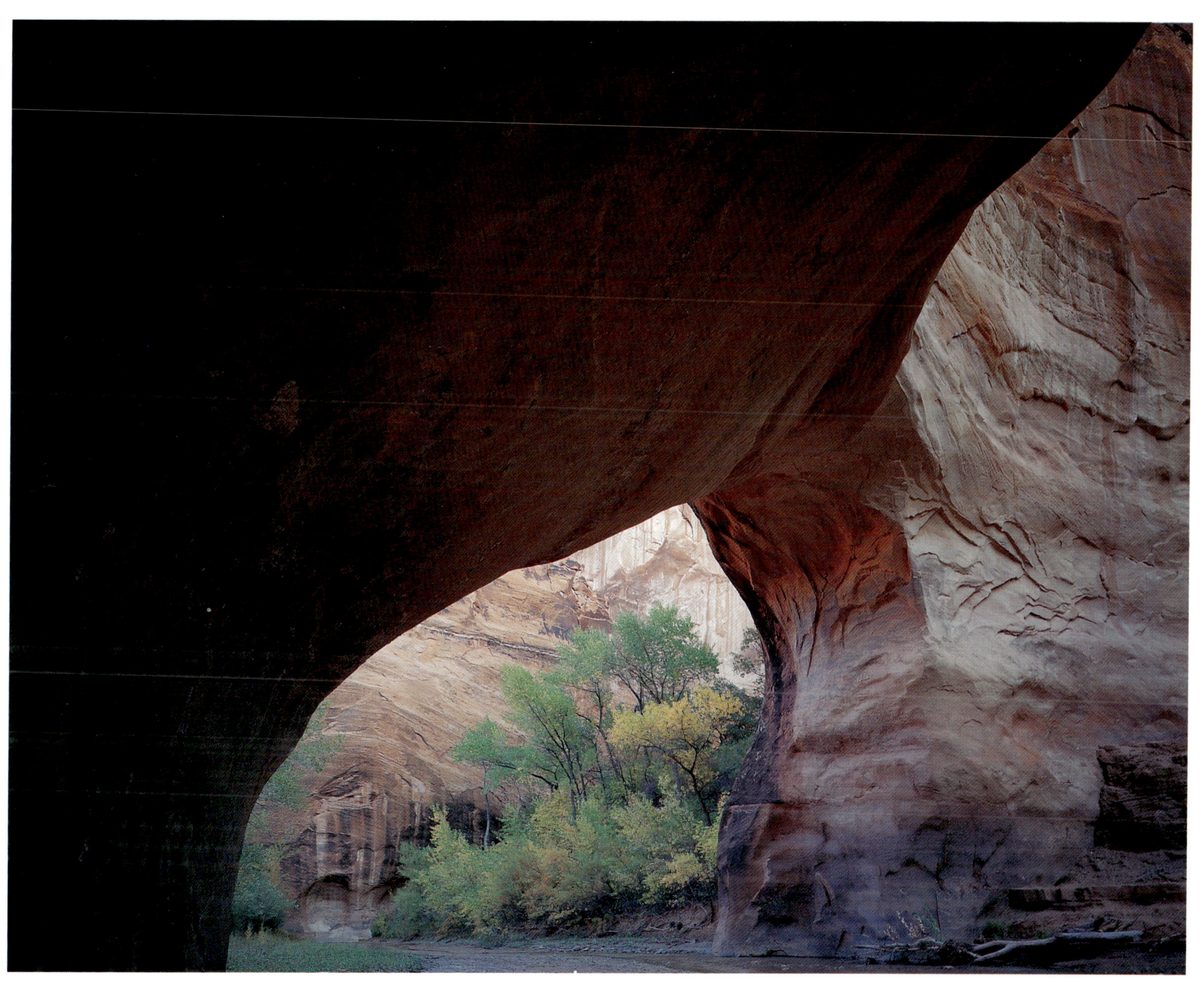

Fall-colored cottonwoods (Populus fremontii) *beneath Coyote Natural Bridge. (Coyote Gulch, Glen Canyon National Recreation Area, Utah)*

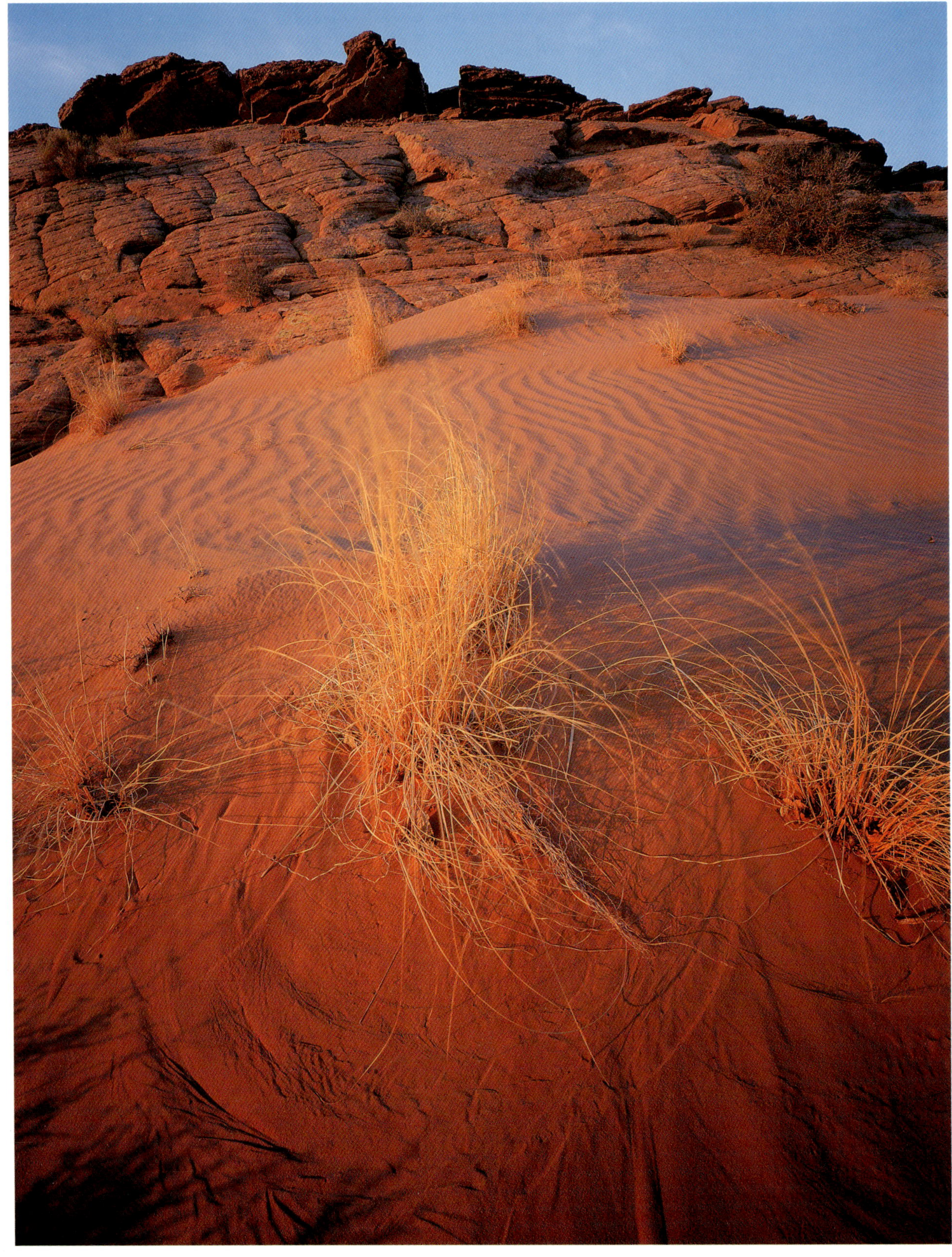

In Dry Fork Coyote Gulch, Indian rice grass (Oryzopsis hymenoides) *grows at the foot of petrified sand dune formations. (Wilderness Study Area, Escalante Resource Area, BLM, Utah)*

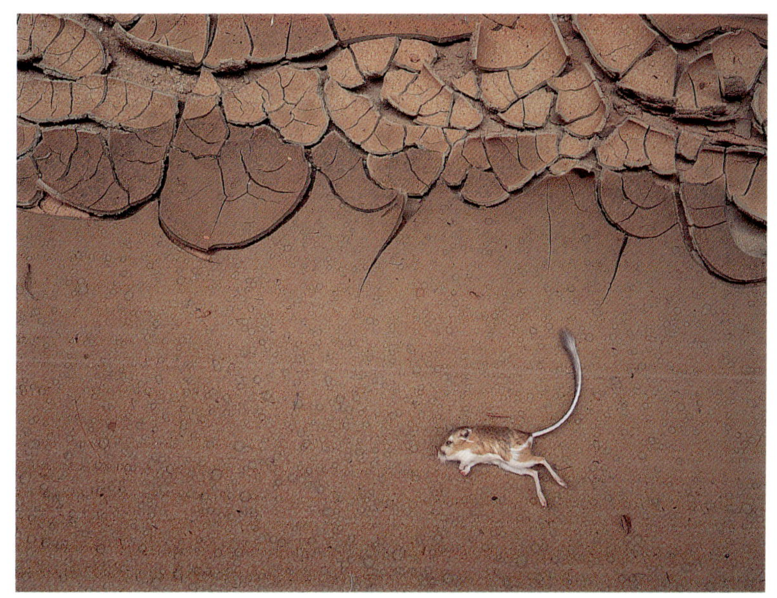

A Kangaroo rat lies dead on the cracked, rain-dotted clay floor of steep-walled Dry Fork Coyote Gulch. (Wilderness Study Area, Escalante Resource Area, BLM, Utah)

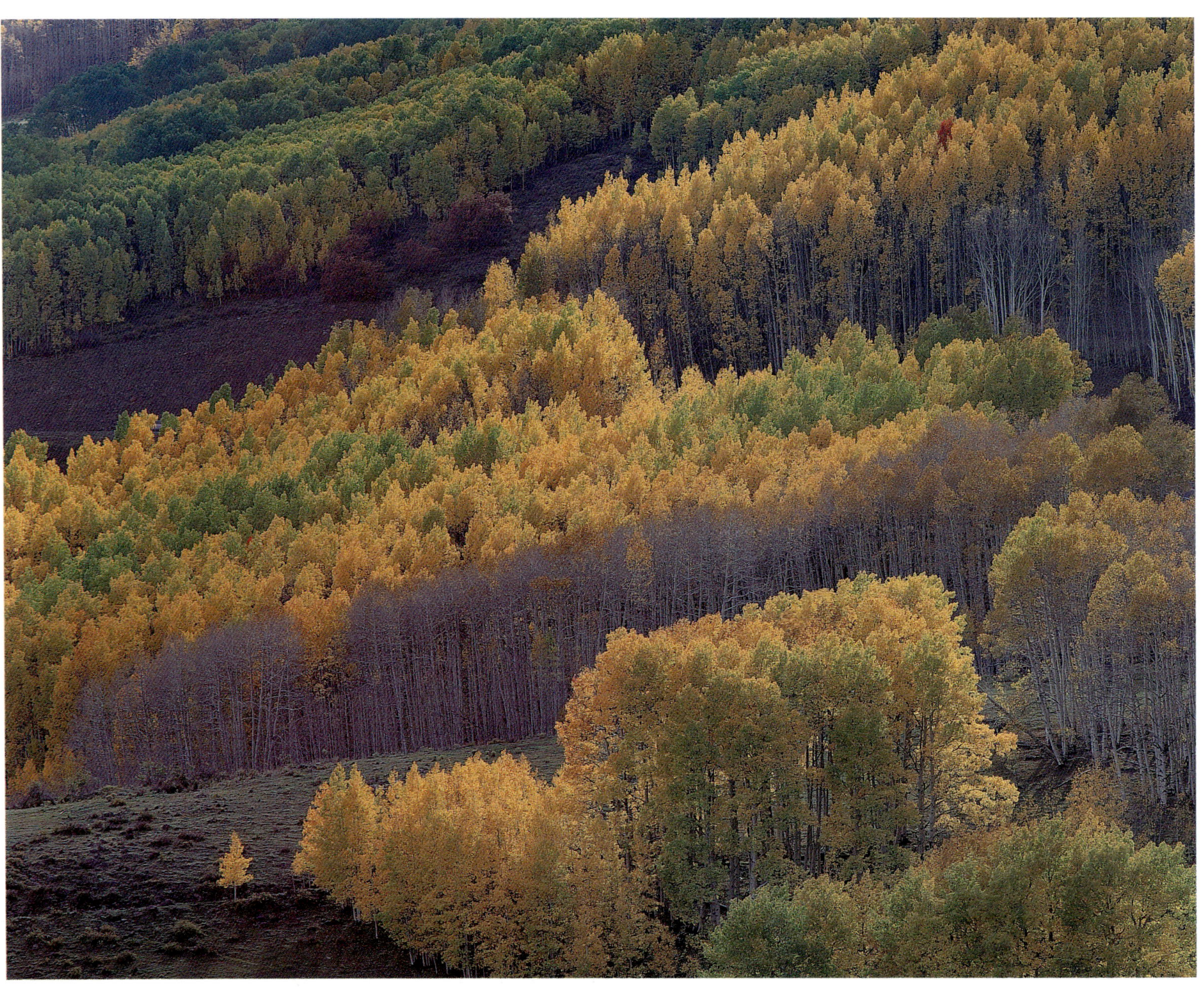

Rows of aspen (Populus tremuloides) *in fall color. (Dixie National Forest, Boulder Mountain, Aquarius Plateau, Utah)*

Lt. George Wheeler, West Point class of '66, is probing the edges of the canyon and plateau country in 1869. He is part of what will become an invasion of scientific Vandals and Visigoths—the expeditions of King, Powell, Hayden, and of course, Wheeler. Everything here is very old, and for the men of science all this antiquity of stone is very new. It will not be until the early 1880s that they will agree on a simple language for their discoveries, the terms we now use, such as Quaternary, Tertiary, Cretaceous, Jurassic, and so forth.

Wheeler is smitten, and in 1871 returns with a passel of geologists and topographers, troops, a photographer, and, of course, a reporter. He is to look for minerals and coal and anything that indicates money can be made. He paddles upstream in the Grand Canyon and travels but fifty-three miles in twenty-four days. His ego is good-sized. Two years after Major Powell and his men have raced down the entire canyon country of the Colorado, Wheeler announces of his midget foray, "The exploration of the Colorado River may now be considered complete."

He is part of a breed. While the Saints are putting their faith and lives on the line to settle their Zion, the scientists are racing each other to cover as much country as possible—hustling like lobbyists on Capitol Hill, pitching their wares in the newspapers, sniping at each other often through the various bureaucracies and agencies that cut their checks.[7] Eventually, he does the Escalante country and in his hands the place becomes blood red with Triassic rock and is surrounded by a ribbon of Jurassic. His map hosts wonderful names—Waterpocket Fold is more brutally seen as Impracticable Ridges, the Kaiparowits is simply Linear Plateau.

What all these ambitious men scamper over is one of the planet's oldest and most stable layer cakes of largely sedimentary rock. They invent a language to record this place and the terms become the keys of the piano on which they play out their geological concertos: on one section of the Colorado, in what is now Grand Canyon National Park, the keys are granite, schists, Bass limestone, Hakatai shale, Shinumo quartzite, Dox formation, Tapeats sandstone, Bright Angel, Muav limestone, Temple Butte, Redwall limestone, Supai formation, Hermit shale, Coconino sandstone, Toroweap formation, Kaibab limestone, Moenkopi formation, Sharump conglomerate, Chinle formation, Wingate formation, Kayenta formation, Navajo sandstone, Carmel formation, Entrada sandstone, Dakota sandstone, Tropic formation, Wahweap sandstone, Kaiparowits formation, Wasatch formation. In other areas there are more keys or fewer keys, but the basic structure remains the same.

The men stand back and look at the work and think they can hear the beginning of time itself, hear the thing ticking in the Precambrian rock too old to imagine and too hard and real to deny. They have named things, and now, they feel, they know them because of these names. This play often consumes their entire lifetime. Wheeler himself is broken physically by the work and retires in 1888.

They are the West we can truly possess. I am holding Wheeler's Escalante country in my hand at this moment. It is labeled Atlas Sheet 59 and captures the dust and heat and thirst of the expeditions of 1872 and 1873 in fine lines, reds, greens, oranges, browns, yellows, and blues. The headwaters of the Paria move very exactly across it and plunge into Paria Canyon like an eighteenth-century minuet. The maps always humor us with their clean order.

But something happens in this hard rock country, even to professional military officers. Clarence Edward Dutton, Captain of Ordnance in the U. S. Army, loses himself in the surveys of the plateau country during the same years Wheeler is roaring about the slick rock. He produces volume after volume of reports, especially his *Tertiary History of the Grand Cañon District.* But science is not enough here. He writes:

> Whatsoever things he had learned to regard as beautiful or noble he would seldom or never see, and whatsoever he might see would appear to him as anything but beautiful and noble. Whatsoever might be bold and striking would at first seem only grotesque. The colors would be the very ones he had learned to shun as tawdry and bizarre. . . . But time would bring a gradual change. Some day he would suddenly become conscious that outlines which at first seemed harsh and trivial have grace and meaning; that forms which seemed grotesque are full of dignity; that magnitudes which had added enormity to coarseness have become replete with strength and majesty; that colors which had been esteemed unrefined, immodest, and glaring, are as expressive, tender, changeful, and capacious of effects as any others. Great innovations, whether in art or literature, in science or in nature, seldom take the world by storm. They must be understood before they can be estimated, and must be cultivated before they can be understood.[8]

We inherit this split tradition of rocks we can name and things in the rocks we can neither comprehend nor ignore. We try poetry, music, photographs, paintings, drugs, alcohol, novels, tracts, guides. Nothing works. There are trilobites swishing in the muck, cycads are growing big as houses; it is all in the rocks, they have left their footprints and we cannot deny them their existence. This is what our brains tell us is embedded in this ancient stone. And more. Dinosaurs, yes dinosaurs also, roaring, making big imprints in the mud, dying and leaving large bones, frightening teeth, huge lizards that make us shudder, all this is there in the rock. The initial humming is, well, we really don't know if it truly is, but we imagine it is some kind of humming right there in the rock, in that oldest Precambrian stuff, the beginning of our world, the original hard rock café, and we reach out and touch it and the rock is hard and cold and yet somehow it scorches us. Then comes . . . well, everything, the entire catalogue of life, something vaster than Noah could imagine, something more diverse and richer than even his fine rainbow could convey. In the rock, you are trudging down a canyon, your hand moves, it touches and rubs . . . the rock. That is what our science tells us, that the book of life is right before our eyes, and recorded in what we determine is cold, unfeeling rock.

We all know this and yet none of us believe it. It is too immense, the units of time are too large, the imagination required exceeds our minds. But it seems to be so. We stand on a cliff and look out at the vista and we think this is beautiful, this is powerful, this is color and texture and a hot wind on our face, and cool water is running silently across the sands in the deep canyon below and we cannot deal with the ideas the rocks are whispering to us. We cannot, no, we really cannot. But here we come face to face with them and so no matter how long we go away, we always come back. If only in our dreams. Huge beasts, dead tens of millions of years, are caterwauling. A mountain bike zips past, the rider encased in modern fabrics. I'm warning you, don't come into this country for a vacation. It will take more. Ask the people rolling those wagons down toward the river to found that new settlement, the people with their children, their ducks, their chickens, their livestock, their musical instruments, their faith. It will take much more than you can give in a vacation. You will be back, I'm sorry, but you will be back. The beasts in the stone will beckon you, the humming of the deepest rock will wake you in the middle of the night no matter how far from this ground you roam. Fair warning has been given. It is not simply one more place. So we must come back for something we truly cannot name and we come back to a place we cannot really describe—even when we have memorized all the strata slowly humming with billions of years of energy under our feet.

Paria Canyon

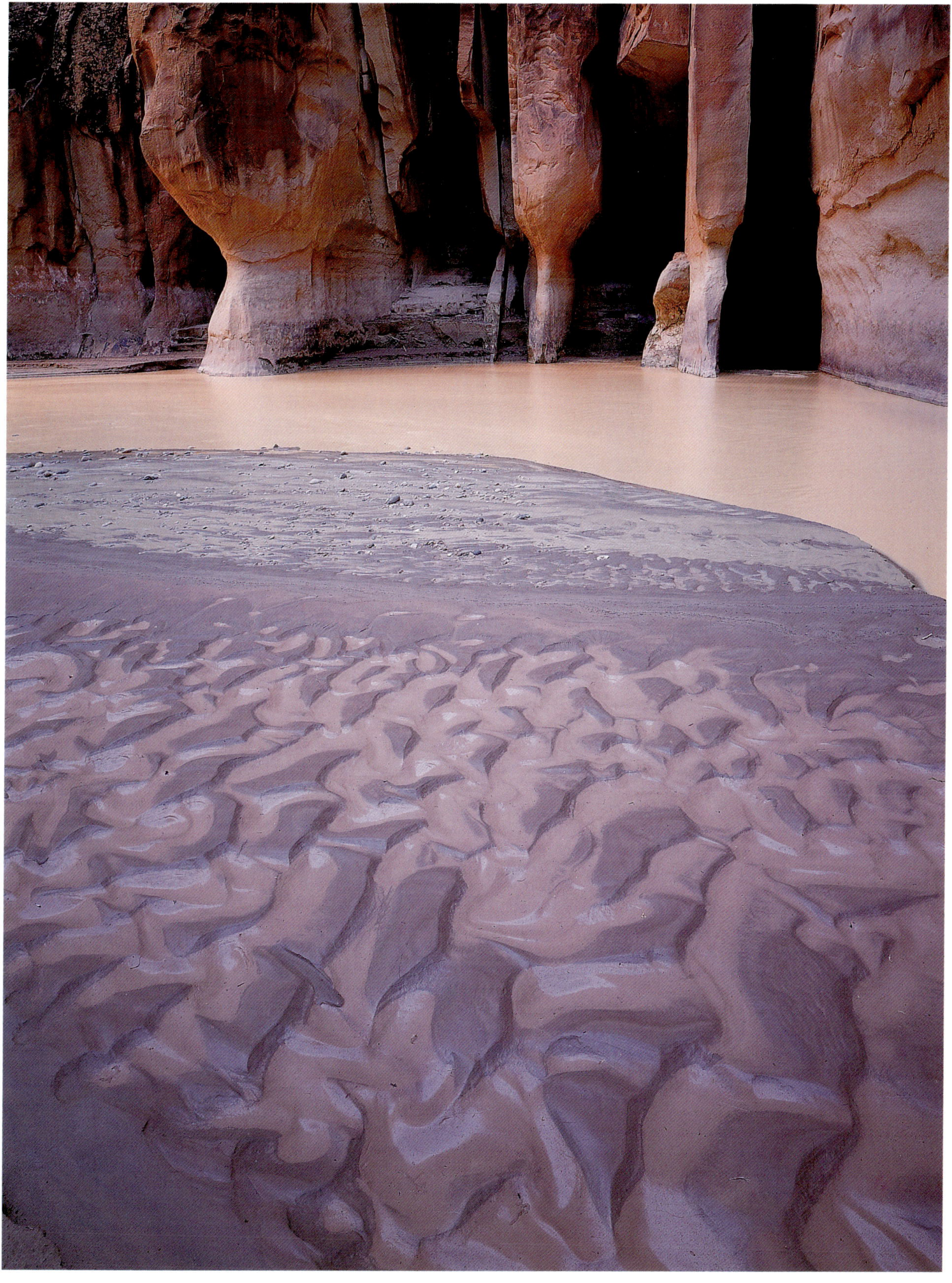

Mixed colored clays and wave patterns seen against the canyon formation. (Paria Canyon, Vermilion Cliffs Wilderness, BLM, Arizona)

Petrified sand dunes with lichen-covered eroded sandstone boulders. (Paria Canyon, Vermilion Cliffs Wilderness, BLM, Utah)

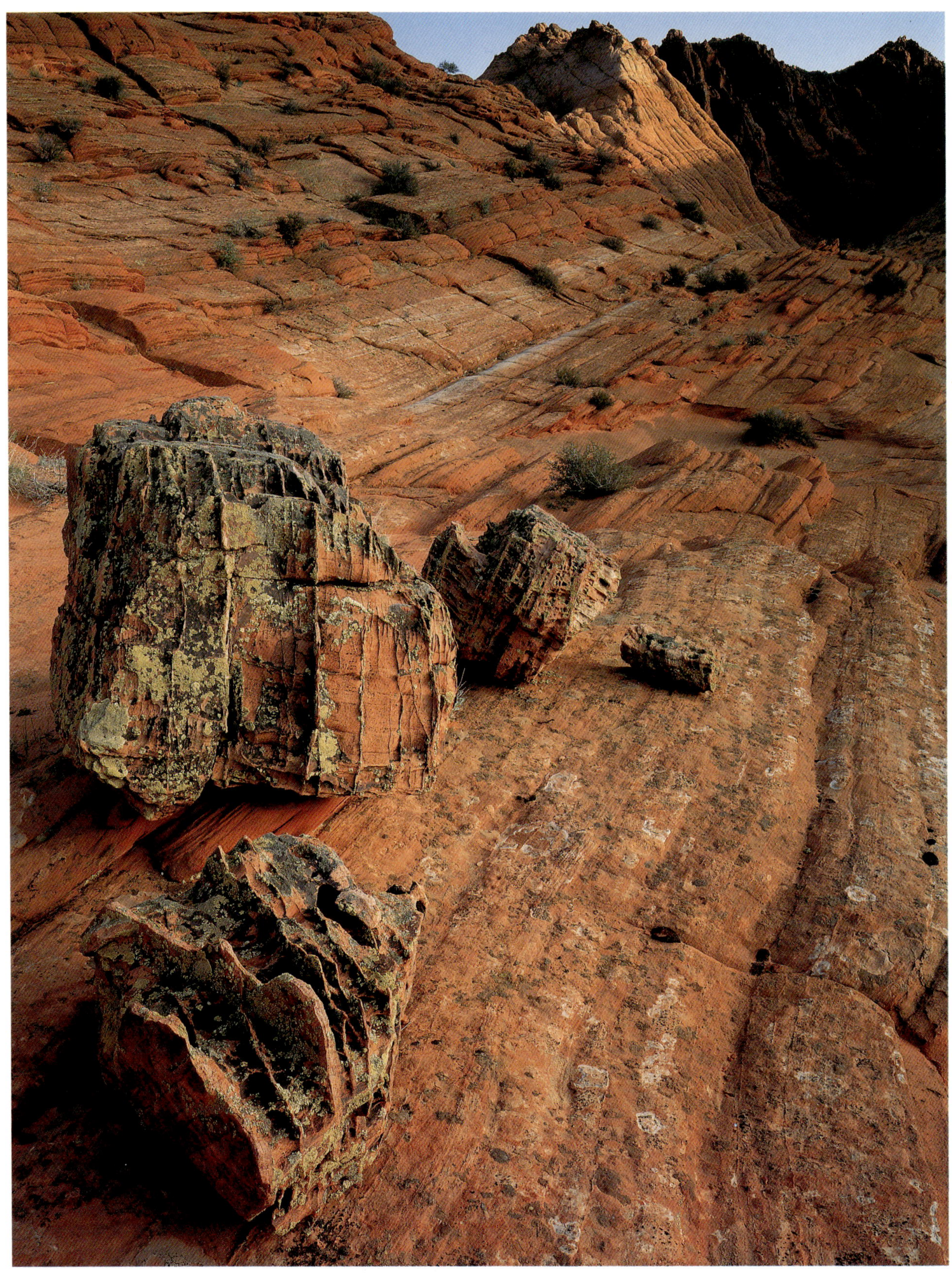

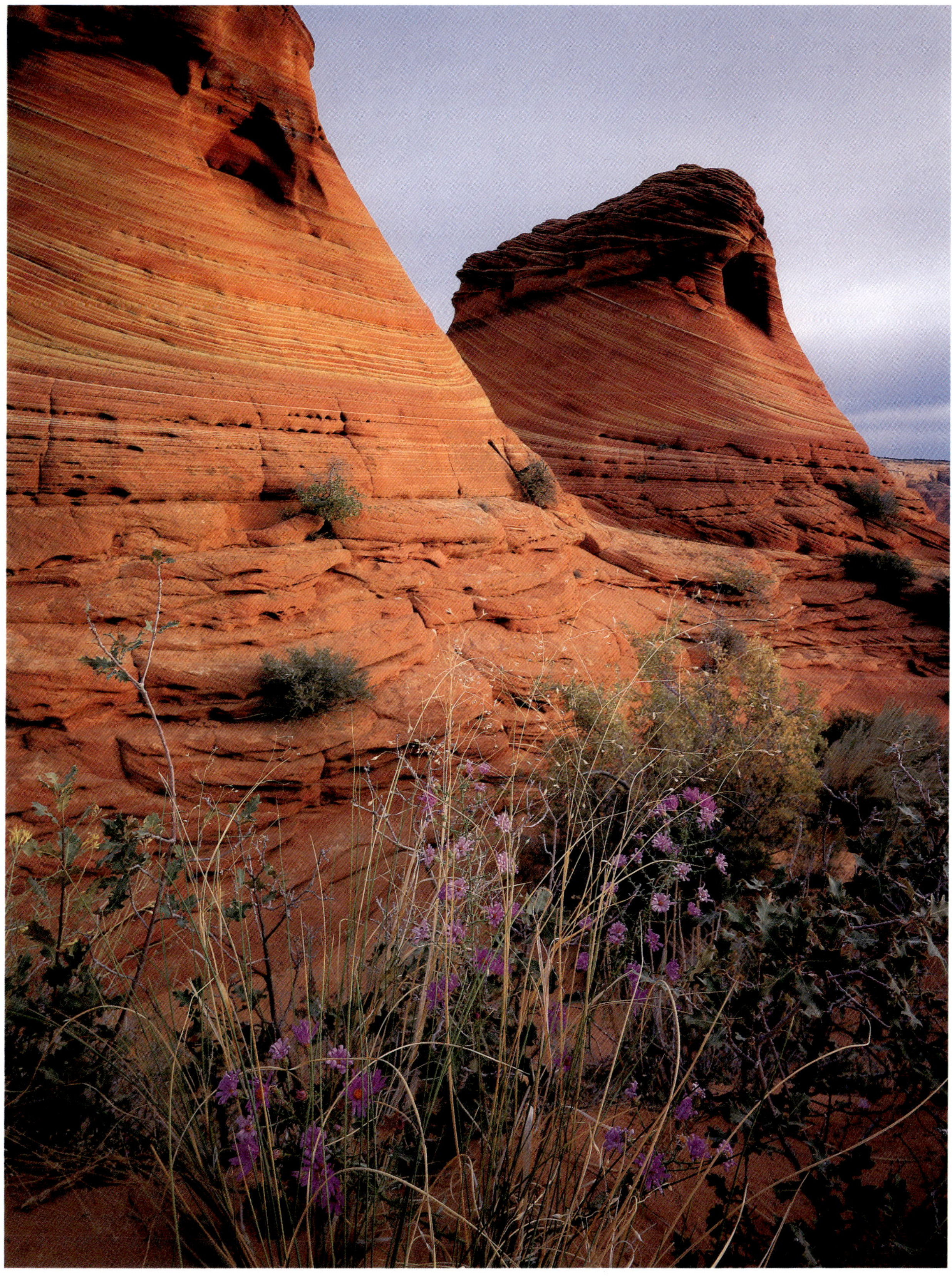

Flowering Machaeranthera canescens *mixes with Indian rice grass* (Oryzopsis hymenoides) *amid petrified sand dunes. (Paria Canyon, Vermilion Cliffs Wilderness, BLM, Arizona)*

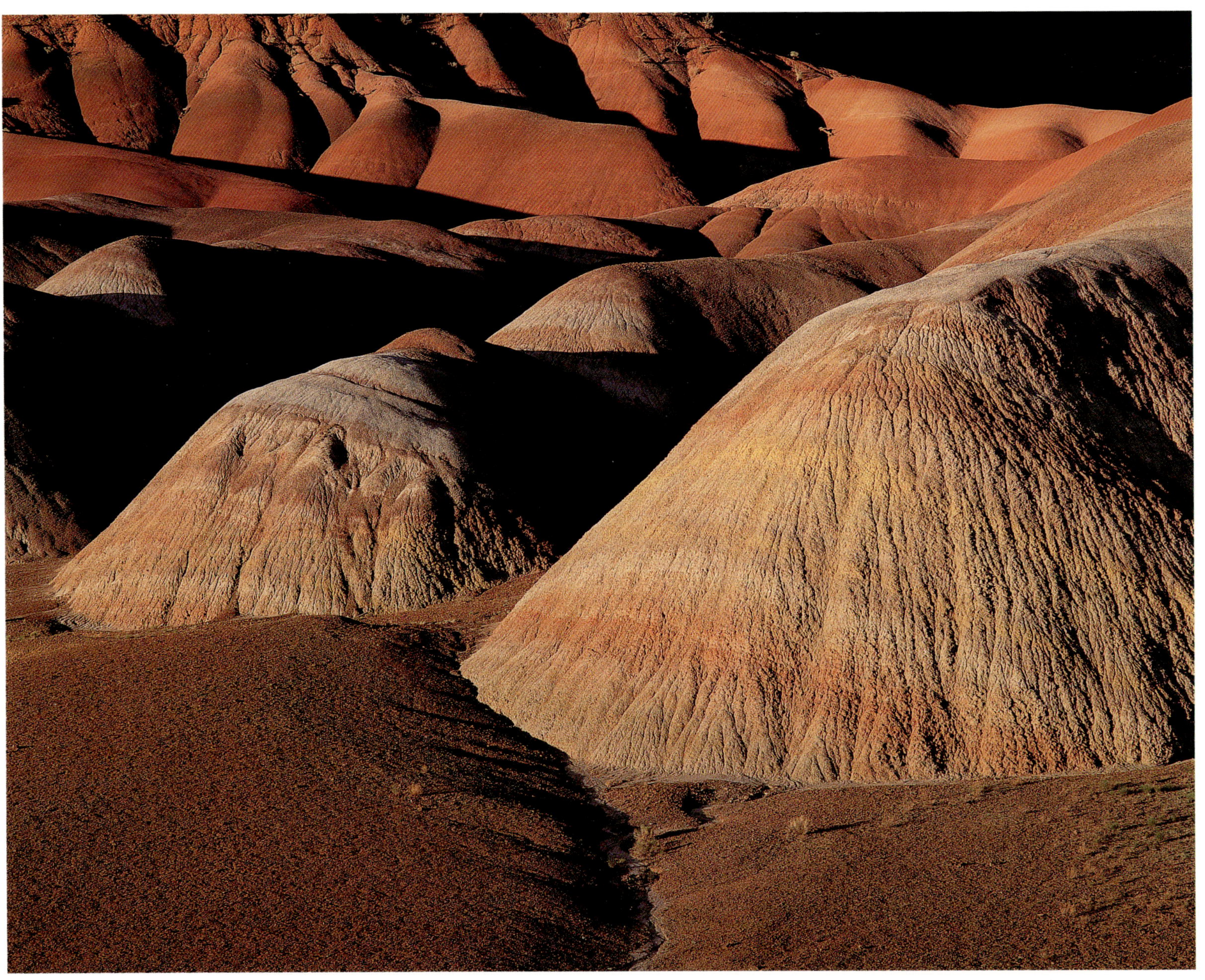

Afternoon light rakes the multicolored Chinle shale mounds. (Paria Canyon, Vermilion Cliffs Wilderness, BLM, Arizona)

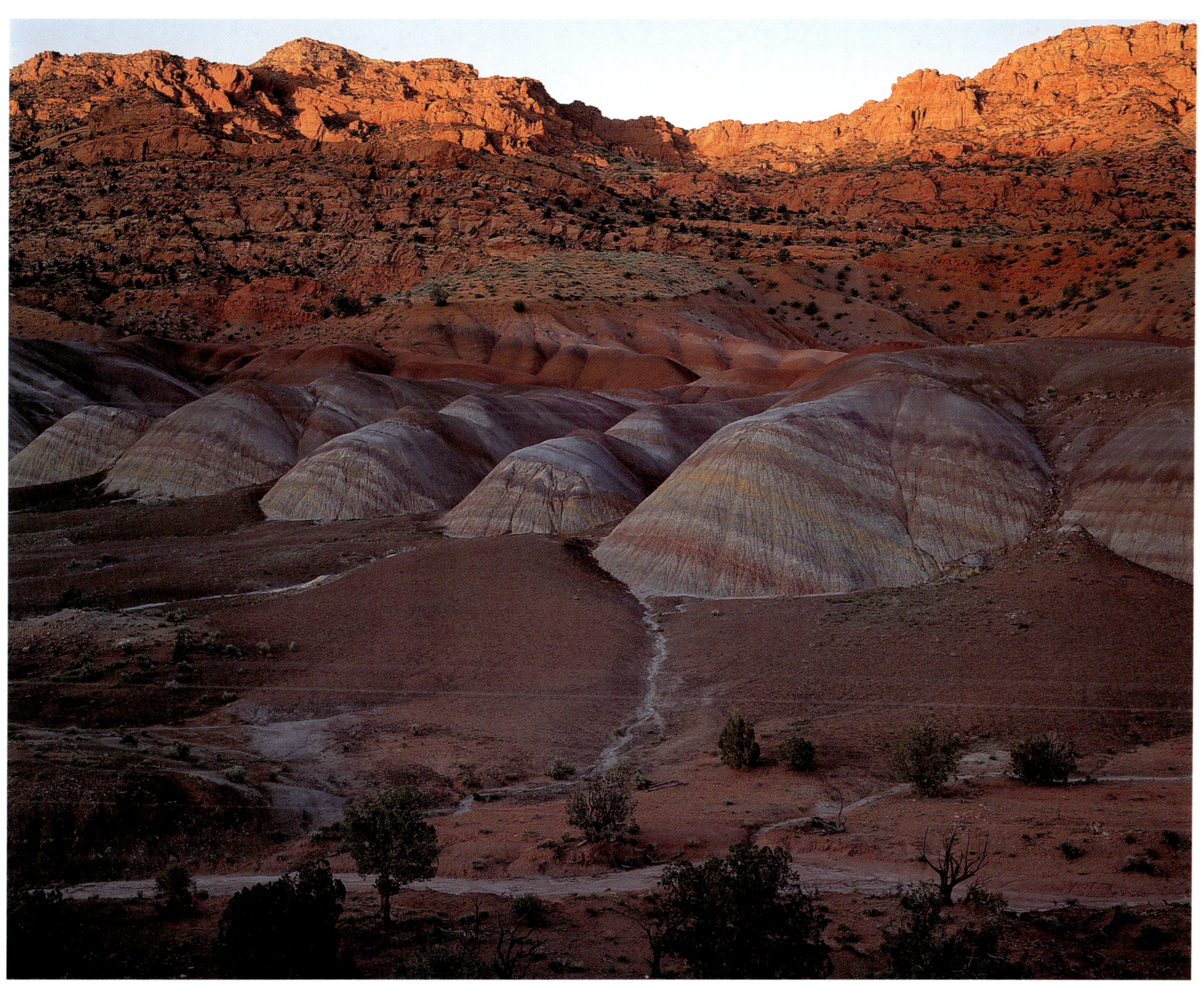

Eroded Chinle shale mounds banded with colors are dotted with junipers beneath the Paria Plateau. (Paria Canyon, Vermilion Cliffs Wilderness, BLM, Arizona)

Manzanita (Arctostaphylos pungens) *grows in the eroded sandstone of petrified sand dune formations. (Paria Canyon, Vermilion Cliffs Wilderness, BLM, Arizona)*

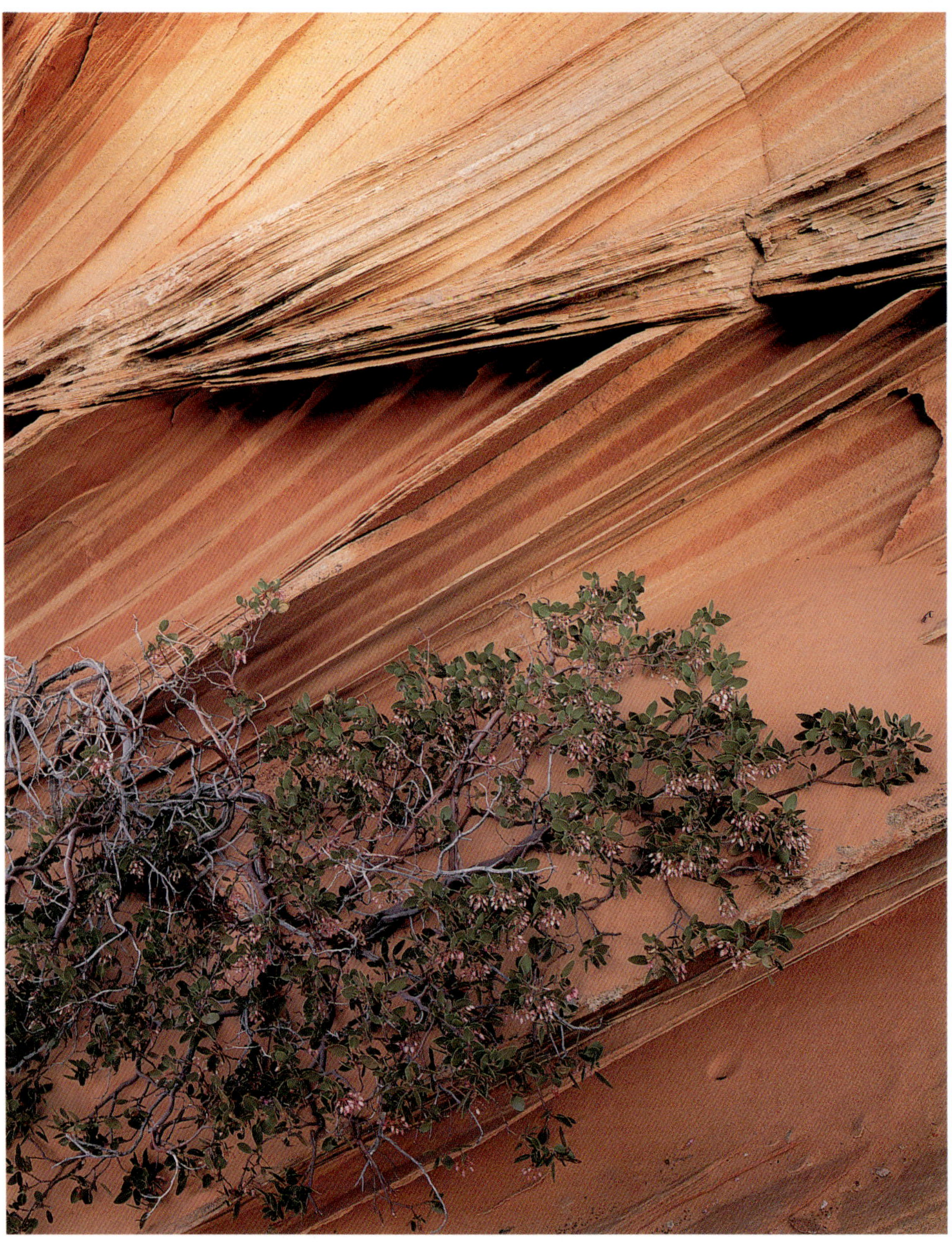

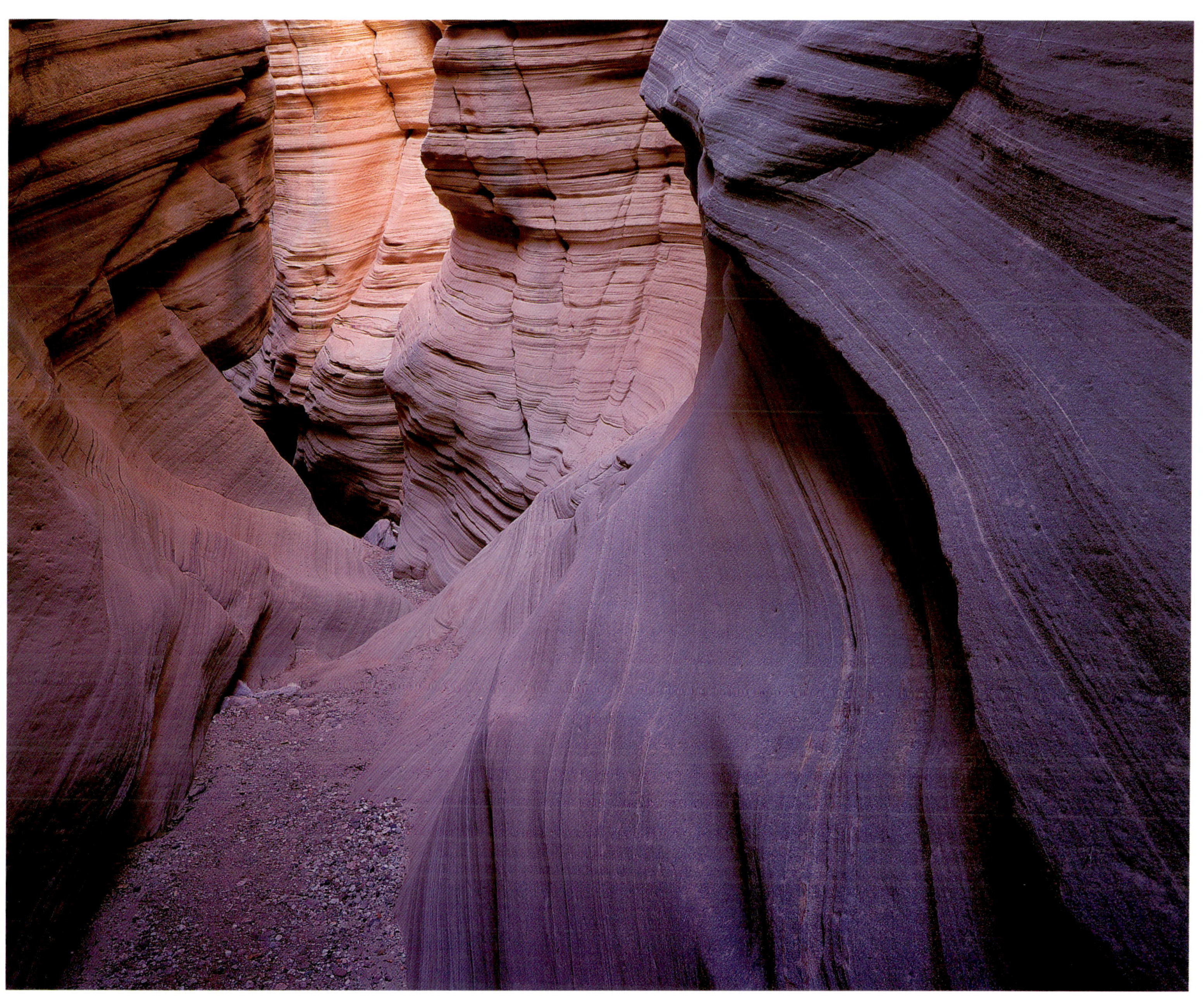

A slot canyon's sculpted walls are of Navajo sandstone. (Round Valley Draw, Paria Canyon complex, BLM, Utah)

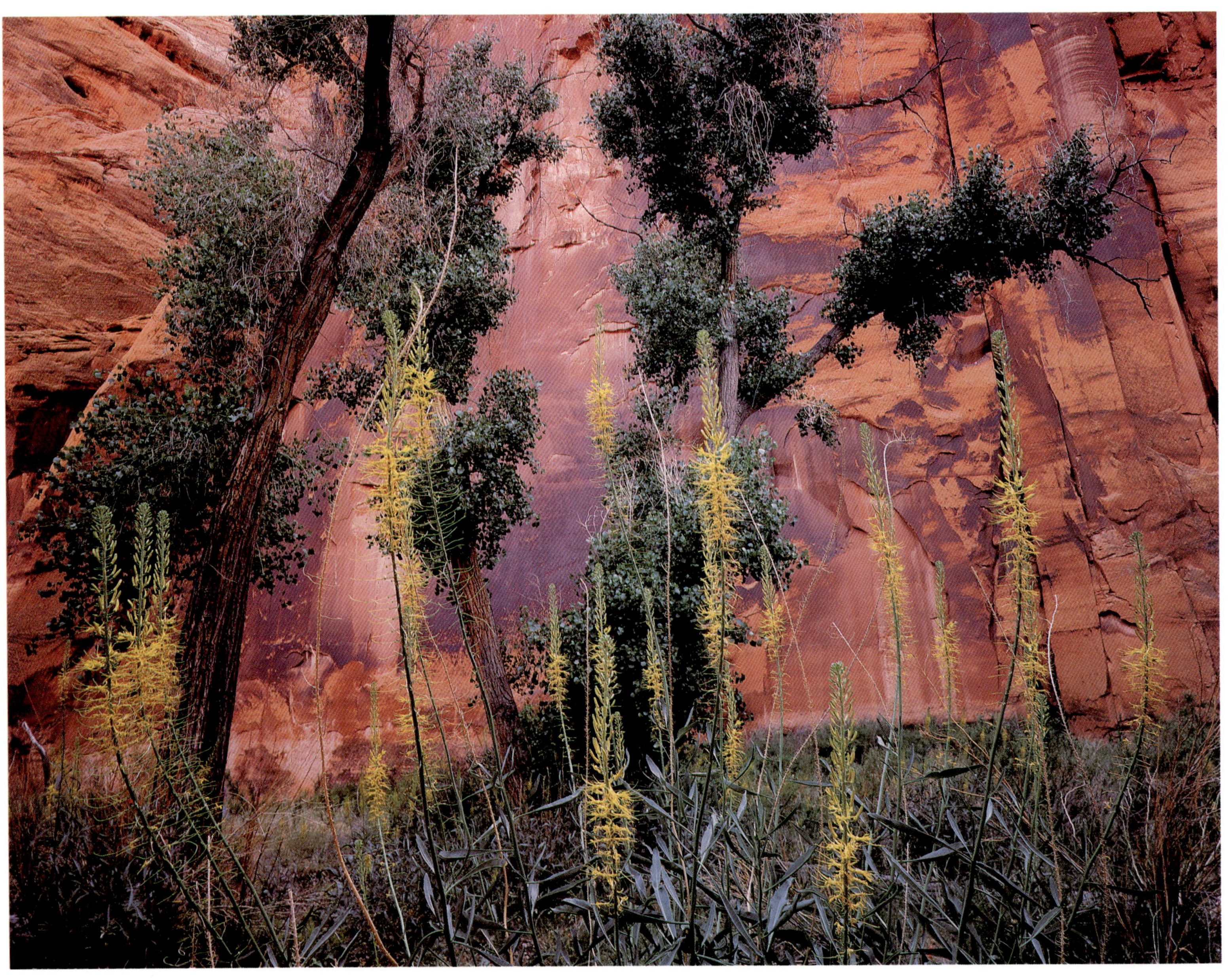

The walls of Paria Canyon appear behind flowering prince's plume (Stanleya pinnata) *and cottonwoods (Populus fremontii). (Paria Canyon, Vermilion Cliffs Wilderness, BLM, Arizona)*

Gambel oaks (Quercus gambelii) *are surrounded by Rydberg sweetpea* (Lathyrus brachycalyx). *(Paria Canyon, Vermilion Cliffs Wilderness, BLM, Utah)*

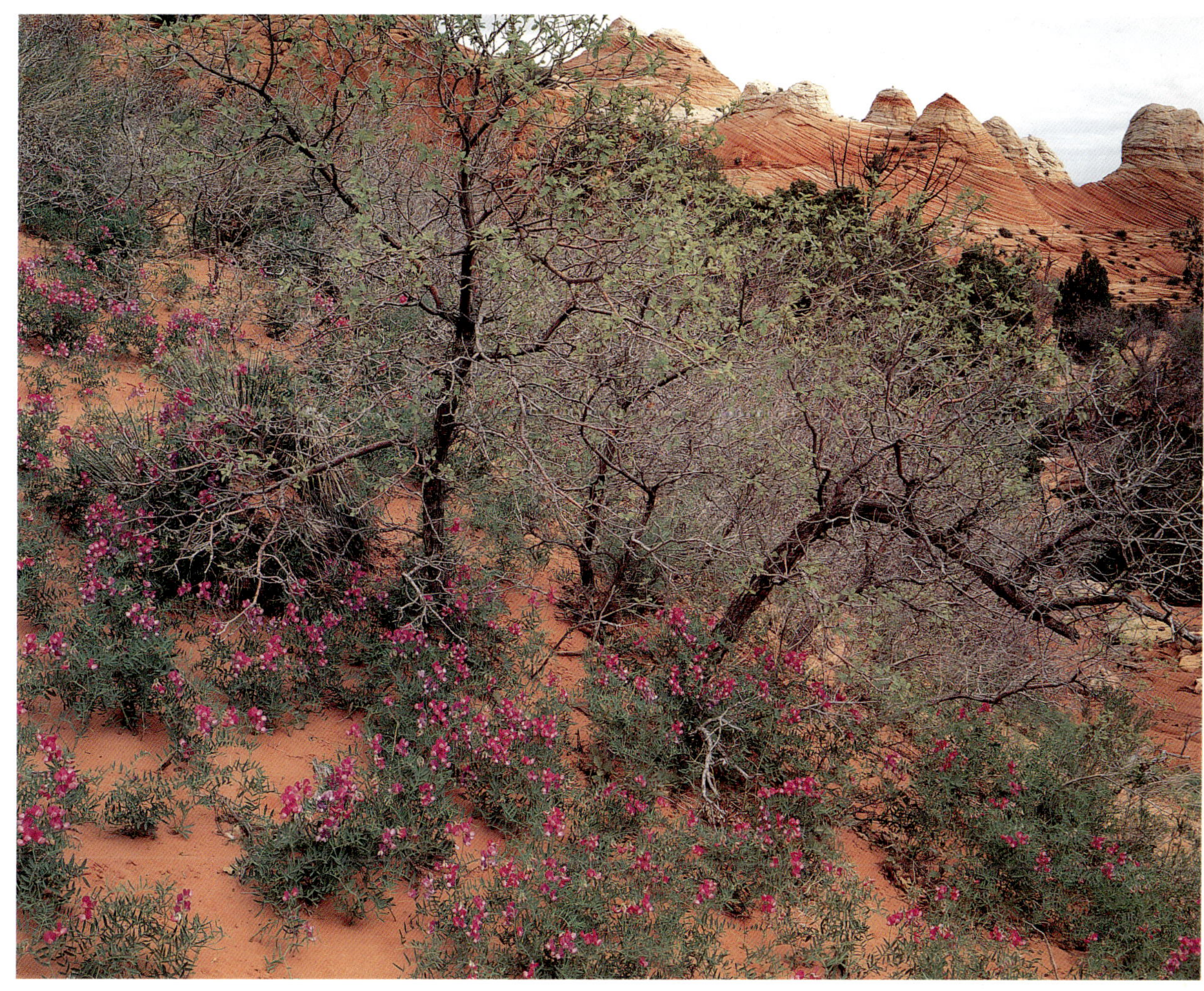

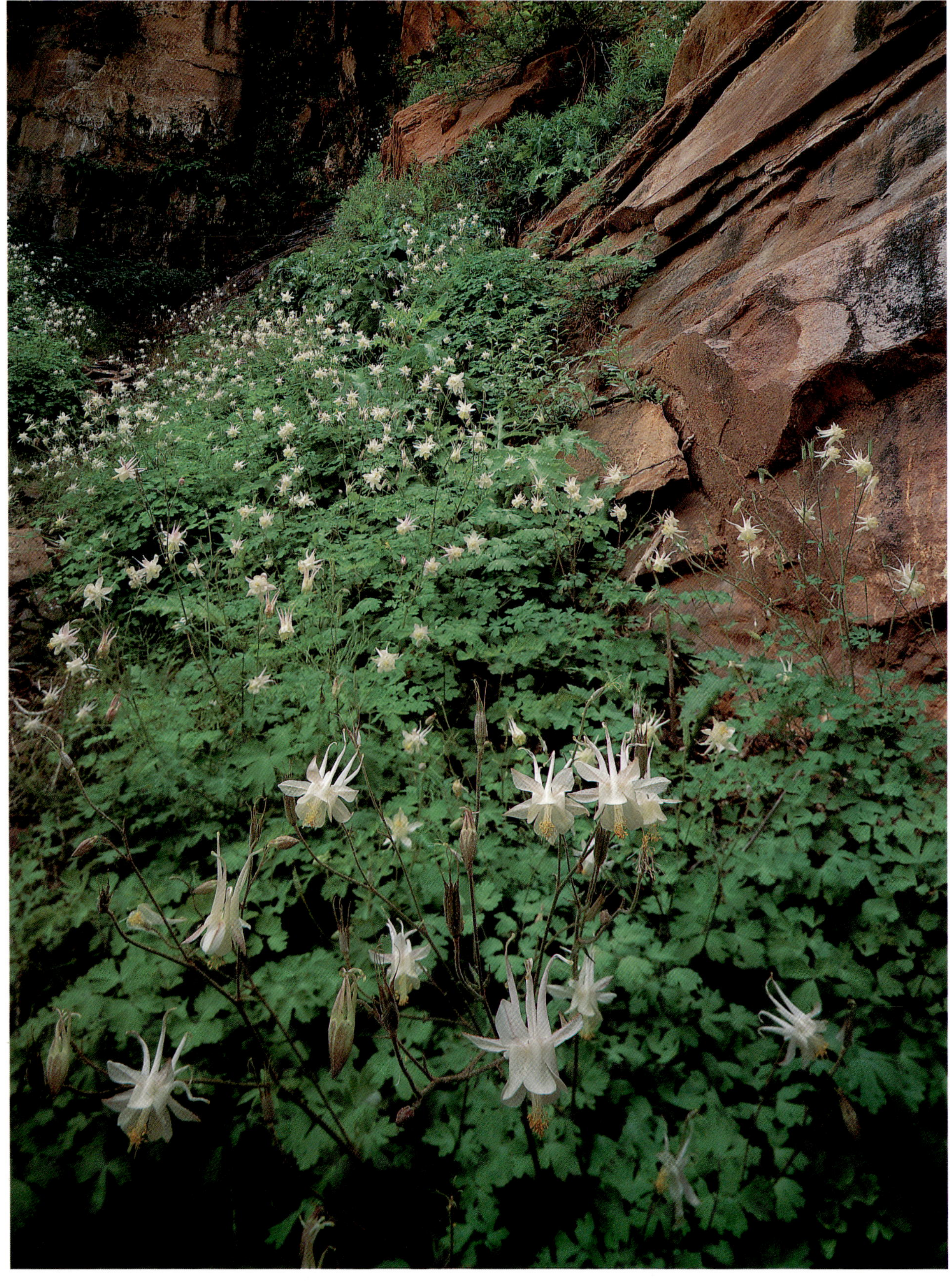

Flowering blue columbines (Aquilegia coerulea) *drink from a Paria Canyon seep. (Paria Canyon, Vermilion Cliffs Wilderness, BLM, Arizona)*

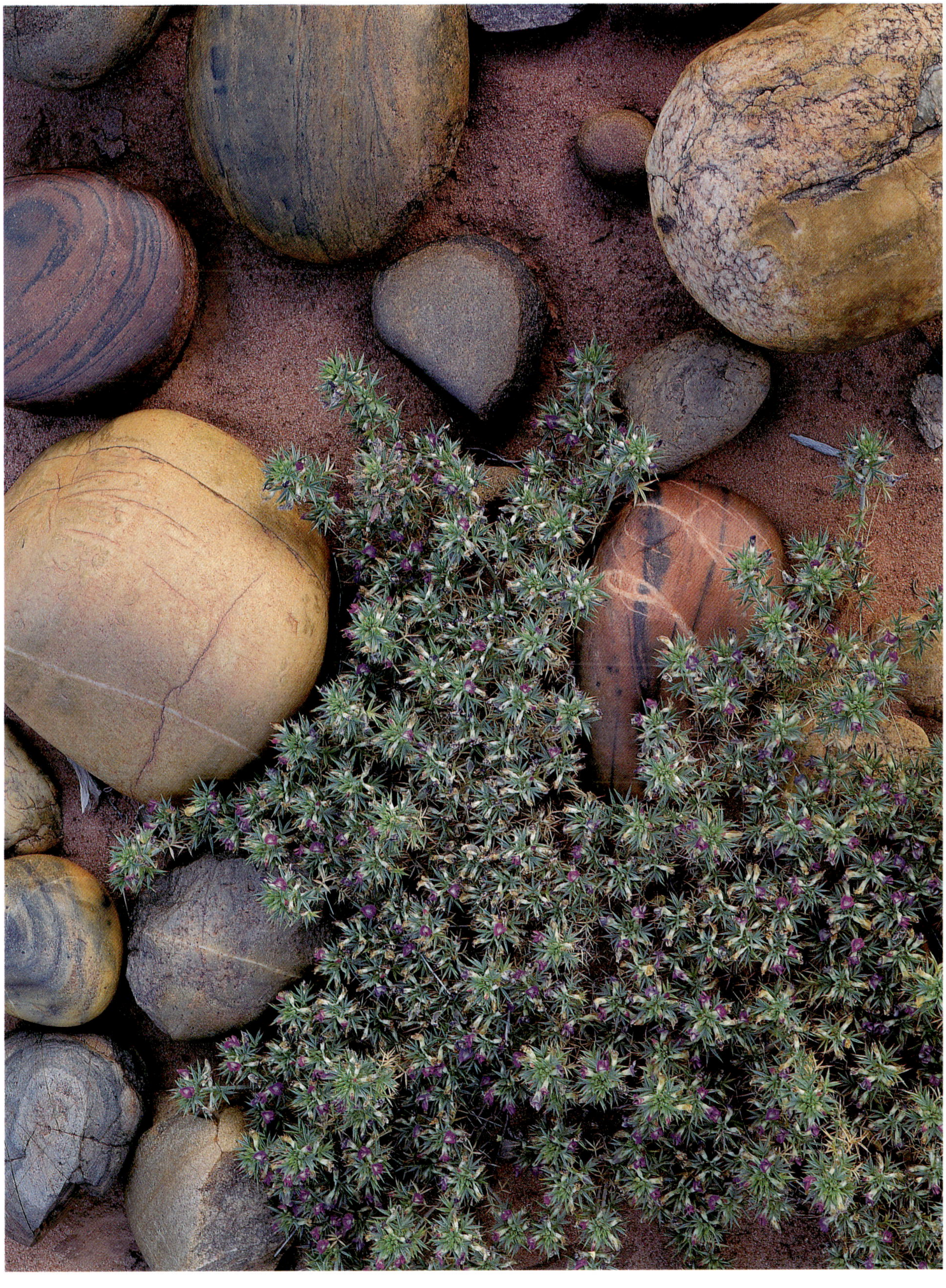

Smooth stream stones and flowering locoweed (Astragalus kentrophyta) *in Paria Canyon. (Paria Canyon, Vermilion Cliffs Wilderness, BLM, Utah)*

Canyon walls are reflected on the mud-patterned shore. (Paria Canyon, Vermilion Cliffs Wilderness, BLM, Arizona)

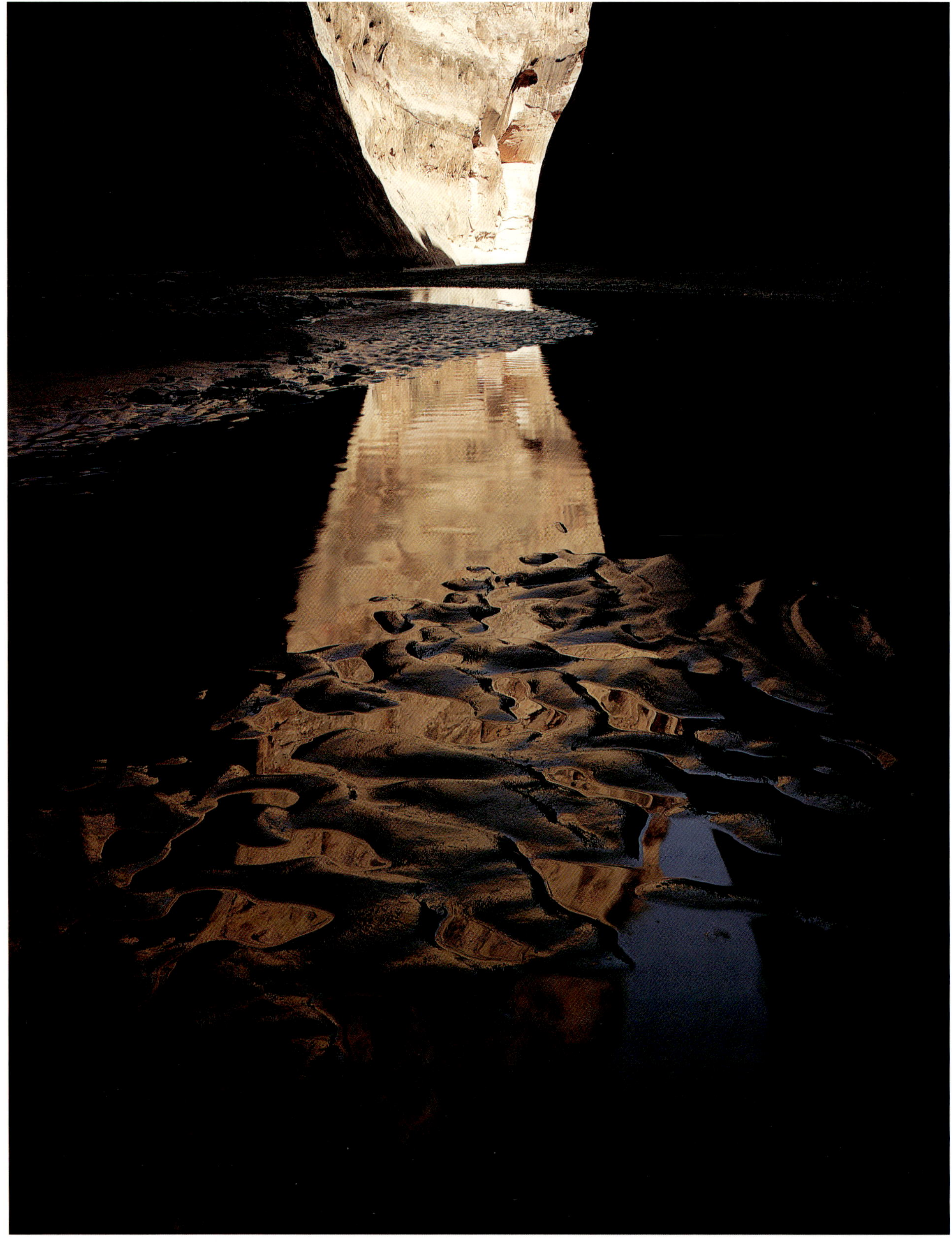

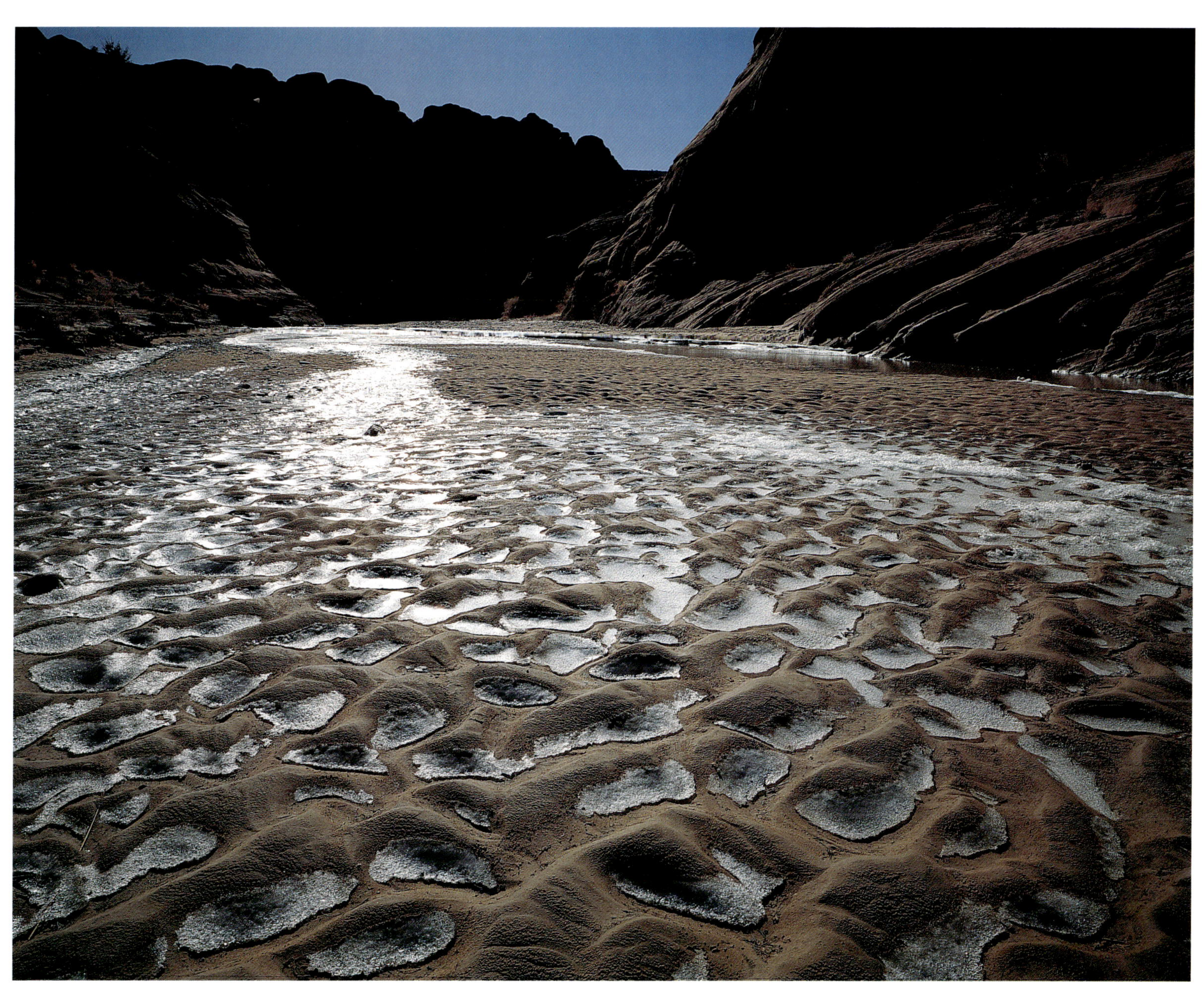

Ice forms in the wave-patterned riverbed of Paria Canyon. (BLM, Utah)

A great horned owl feather lies on the patterned shore of Paria Canyon. (Paria Canyon, Vermilion Cliffs Wilderness, BLM, Utah)

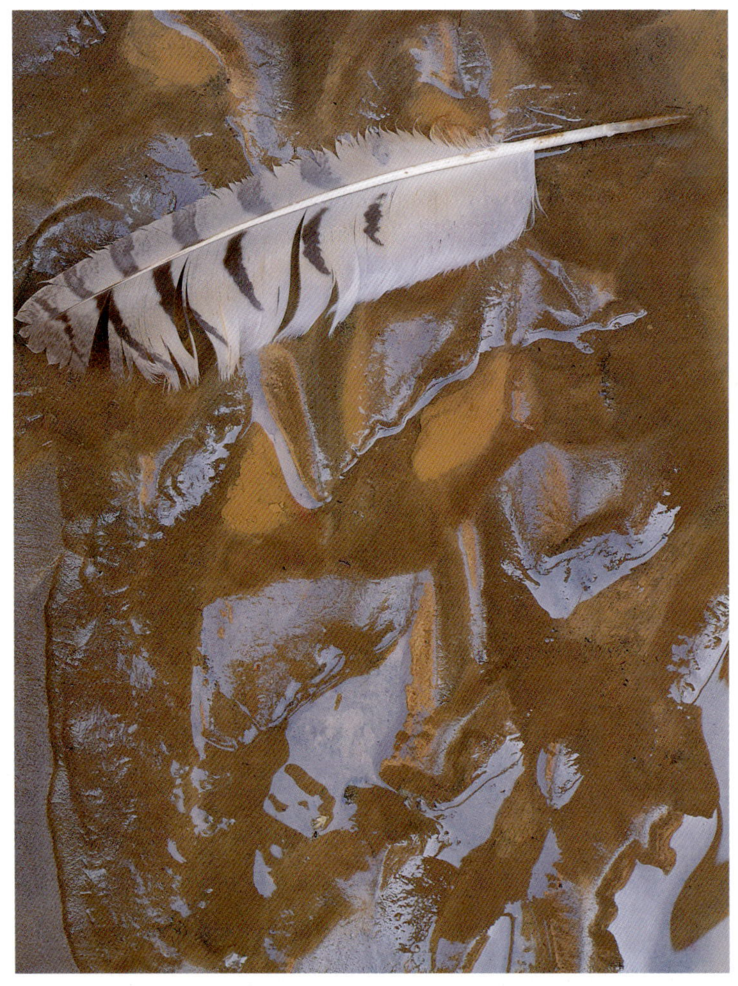

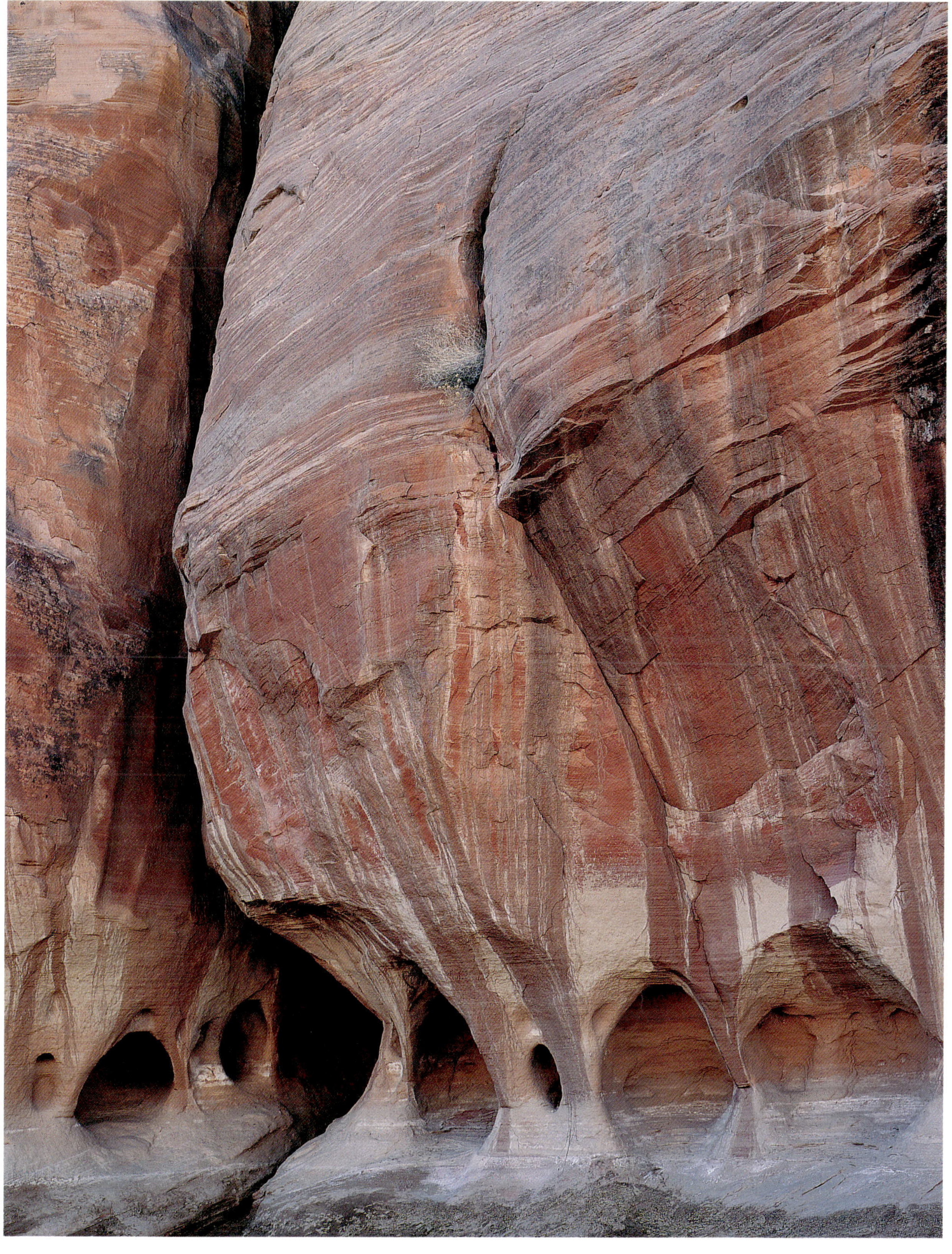

The eroded sandstone walls of Paria Canyon's narrows. (Utah)

He dreams hard dreams. John Doyle Lee is forty-six years old. He stands five feet seven inches, has blue eyes and a ruddy complexion. He can be a sociable man and has been known to feed at his table one hundred to four hundred people. Or he can be a very hard man, hard with the dollar, hard with his brethren. His nights are flooded with images. He has much to draw on. He is that creature we call a natural leader. He is also a man of deep faith and sees the world as a series of events in which, through observation and prayer, he comes to know God and God's ways. Once Brigham Young tells him of a prophecy that ran "in the name of Israel's God, this man Lee who now is some much spoken evil of, will yet destroy and trample under his feet, and walk over their graves, those that would destroy him."[9] He is an adopted son of Brigham Young and in many ways his right-hand man—the person always selected for the hard jobs, the dangerous jobs. And he is a builder, a man who leaves his mark all over southern Utah and his name on one of the most lonely places in the United States.

He was part of the Mormon migration to Missouri that was later mostly slaughtered, the survivors driven out. He helped build up Nauvoo in Illinois and served as a bodyguard for the Prophet Joseph Smith. He was a key figure in the vast migration to Utah, in which the Saints fled beyond the boundaries of the United States seeking Zion. And then, after only two years, the Mexican War caused huge changes on the map and the Saints were suddenly within the borders of the United States once again. He is a man of many wives—he had ten already when he left Nauvoo.

On this Saturday, February 19, 1858, he has a lot to remember and a lot to forget. It is just five months since that September afternoon in Mountain Meadows. Then, in a few minutes, he had witnessed the slaughter of an entire wagon train of men, women, and children. Afterward he had mounted his horse and ridden down into town where one of his wives awaited him. He had no words to describe what he had participated in and she had no questions. She had already tended to the few children who had not been murdered. Louisa, the three-year-old with the arm shattered by a bullet, could not sleep. But John Doyle Lee was exhausted that September night. A new moon rode in the sky and the air sagged with the scent of balsam and sage as he climbed onto a haystack and fell into his dreams. Before leaving the killing ground of Mountain Meadows, he had told his men that they had done their duty in defending Zion and their families. He had ordered them to stay on the ground until morning so that they could look to the burying.

Now on this February day he recalls a vision that had come to him twelve months before. The memory of that vision is like a stone pressing down on his mind. Back then, he had been riding along thinking about the weakness of human beings, of their many failings, when this vision came and filled his "Heart with Joy and thanksgiving."[10] In his vision, he enters a fort and a bed is pointed out to him. All the inhabitants of the community are sprawled out in the chamber. Lee goes to the north side of the room. It is night. Soon, Lee notices a man passing by at the feet of the sleepers. He is making notes with a silver pencil, and he moves from south to north. Lee hears the man say repeatedly, "You are not all Saints that are reposing here. Some are Dogs." Lee argues with the man saying, "Not so. They are all Saints." He knows better, but "felt ashamed to acknowledge or expose the weakness of My Brethren." The man walks over to Lee and keeps calmly making notes. Suddenly the sleeper next to Lee stretches and his hands and feet slip out from beneath the covers and "to my astonishment they were Dogs' feet, ears, and Head." The notetaker asks Lee if he is convinced now, and Lee looks around and realizes that most of his companions are dogs. Suddenly, the stranger drops his pencil and jewels pour out of it. Lee leaps up and says, "I will no longer be identified with Dogs." Night begins to give way to dawn, and when Lee tries to help gather up the jewels they are nowhere to be found. He discovers a table covered with fruit and vegetables and suddenly new strangers explain the meaning of each item and the moral value of each Saint, and Lee is suddenly cast back into the days of last summer and the events that led up to that day at Mountain Meadows. He now realizes that "my denying that there was Dogs in bed with Me was my covering up the faults of the people; the light was the reformation."

The vision is a comfort for John Doyle Lee. It helps stop the screams that still ring in his memory. But all is not sadness in his life. In the summer of 1858, he has twelve men busy painting, papering, and putting down new carpet in his house. He has made three hundred gallons of malt beer, slaughtered two cows and two goats, and for days his womenfolk have been baking. He hosts a gala for the Fourth of July. Then, in August, he attends a meeting with the highest officials of his church. The talk goes on for two days, but he senses something different in the air. He does not know that some of the church leaders have come to the meeting by way of Mountain Meadows. There they saw bones scattered about the ground, clumps of human hair clinging to bushes, here and there fragments of faded sunbonnets.

Still, all seems well for him. He and his many wives, his large flock of children all prosper. He is busy building up his properties and serving his church. His diary is studded with his trials and his successes. "A darker time," he notes in 1868, "I have not seen for many a year to sustain a large Family with Bread. About 50 Mouthes to fill daily—yet my trust is in god. I have all my Bread, Meat, & vegitables to bye till after Harvest & but few cows to give us Milk yet. My Promise is by the Prophets of the Lord, my own & others, that I shall never lack, nor my Own beg Bread & in times of Famine I shall be fed & that Means Shall flow to me from unexpected quarters."[11] He has his dreams also and he carefully records his dreams but these visions fail to tell him one important thing. He has already begun his journey into the heart of stone.

I AM COMING down Paria Canyon in January. The water in the stream is steady at thirty-three degrees and here and there ice covers the channel. The cliffs are massive, and sunlight barely penetrates the slot canyon. Quicksand sucks at my feet, and now and again the heavy pack drives me in almost to my waist and I must flop out on top of the muck and do a kind of frog stroke to firmer ground. The walk goes on for days and days, fifty or sixty miles of cold and shadow and beauty. The dive of the Buckskin watercourse corkscrews in from the right, and the walls are barely a shoulder width apart, the light is a forgotten thing. Even in the main cut of the Paria there are places only as wide as a car. Here the world seems to disappear, or at least the world that normally commands my attention. I am deep into the rock.

At night it drops to ten degrees and I huddle in my sleeping bag with fingerless gloves and read Carl Sandburg's massive biography of Abraham Lincoln by the flicker of a small candle lantern. The January nights are long and I read late as the cold settles like Jell-O on the canyon floor. Lincoln is in the middle of bloody civil war, and he is laying down riffs as hard as the stone walls that frame my canyon bedroom: "A nation may be said to consist of its territory, its people and its laws. The territory is the only part which is of certain durability." Sandburg underscores this point with quick hammer blows of prose—"Laws change; people die, the land remains."[12] I am living in a wonderful time warp, what with rock as old as the planet scattered about, Lincoln booming up from the page with his notion of the necessity of the Union, and old John Lee moving down the canyon in flight from that very concept of Union, the same flight that had driven the Saints across the plains in their desperate desire to found Zion. Lee is moving his herd of cattle down this narrow canyon, roping them and pulling them out of the quicksand. He is headed for a place no one really knows, a place one of his wives will call Lonely Dell. He is literally fleeing the United States of America, and here in the stone wilderness he is confident he can achieve his goal. My ambition is more modest: to simply taste a little of what he must have felt as he moved his world during that long ago winter into this last safe place. Of course, I can never duplicate what is in his heart, but I can set my mind free to visit and imagine his world.

In the morning, I load up my pack and wade into the freezing water. The ice breaks and I begin my daily bouts of falling into the water and then leaping up again with an instant chill coursing through my body. I round a bend and a huge slice of sandstone lies on the ground. I look up and see a fresh scar where it has calved off the cliff. Everything in the area is covered with a fine pink dust—the thing has just happened. I have heard nothing but the dust tells me little time could have passed. So I stand there staring at the eternity of the rock face, trying to imagine the lightning movement of the plunge of the gigantic slab, and am befuddled by the quick moves of something as eternal as stone itself. Usually, I see the various strata and tick off their names and think that the eons they express are so far beyond my honest comprehension that all this geology I periodically gorge on is a hopeless venture, just as theology makes sense for a moment, yet does not seem to bring me closer to the mysterious face of God. And then I think that my effort to chase down and capture John D. Lee in his flight of a century or more ago is not so impossible. Here time means little, and time means everything. Here there is lots of time—just look at any canyon wall.

THEY ARE TEN and they leave Santa Fe on July 29, 1776, seeking a quick route to the West Coast. They have a herd of horses and a string of pack animals and a deep need to avoid two things: the unfriendly Native Americans of what is now northern Arizona and the endless canyons of what is now southern Utah. So they go north into Colorado, spin eastward toward the Great Basin just below the valley of the Great Salt Lake, and then head southwest in the hopes of reaching Monterey, California. Father Francisco Atanasio Dominguez is in charge. Don Bernardo Miera y Pacheco, who travels with an astrolabe, compass, and quadrant, is the official mapmaker. Father Silvestre Velez de Escalante keeps the journal, and so he becomes the one person history truly remembers.

Their journey inscribes a huge circle and no expeditions that come after them follow up their path. They are wise travelers and have little trouble with the various tribes they encounter. The land is another matter. They periodically starve, almost die of thirst, and almost freeze to death. The fathers promise various tribes encountered on the trail that priests will come within a year to teach them Christianity and modern farming. But the Spanish crown never follows up on the expedition's report. They are like stones cast upon the waters of a pond; when the ripples die down there is no evidence of their passage. They are victims of court politics, the intrigues of various religious orders, the warlike tendencies of various tribes—but in the end their ambitions come to naught because of the stone.

The first turning point comes for them on October 11. They are camped in southwestern Utah on the edge of what today is called the Escalante desert. For days a cold north wind has worn them down. When they look south and west into the dry country separating them from the California coast, they stare at five hundred miles of cold, thirst, and hunger. So they balk and talk and bicker. Miera wishes to press on—he dreams of fame for carving out a route to fabled Monterey and the California missions. The priests disagree—they see death waiting out in the dry ground. So they pray and then cast lots and Miera loses. They then turn back and angle south and east, seeking a way to needle through the Colorado Plateau and reach their homes. For two and a half weeks they stagger through the dry mesas and imprisoning cliffs until, on October 26, they come to the mouth of the Paria River, where it flows into the Colorado: what is now called Lee's Ferry. Twice they have had to kill some of their horses in order to avoid starvation.

"This afternoon," Escalante writes, "we decided to find out if after crossing the river we could continue from here toward the southeast or east."[13] Two of the men who are good swimmers strip off their clothing and, with their garments on their heads, plunge into the Colorado. They barely make it across. Midstream their clothes vanish into the turbulence of the river. Naked and weakened, they have little energy for exploring the far shore. After resting and getting their wind back, they return naked and cold to their companions.

Escalante names this camp *Salsipuedes,* Get Out If You Can. They build a raft, but the river is too deep for poling across. In desperation, the fathers dispatch two of the men up the Paria with instructions to find some escape. The party waits a day, and then slaughters another horse. Four days later, the men return. They have technically found an escape route, one that Escalante describes as "extremely difficult stretches and most dangerous ledges, and at the very last impassable." When they climb up out of the canyon, they are standing 1,700 feet above the Colorado River. For two days they struggle across a mesa. They send out two men to find a way down. The next day, as the party waits, they have nothing to eat but cactus and wild berries. One man returns and reports there is no way down. They spin toward the northeast, and then in the night it rains heavily and later the rain turns to snow. After three days the second man turns up, without his britches. They press on into the face of a blizzard. With morning their luck changes and they discover a steep but possible route back to the river.

The horses make it across, and this becomes the Crossing of the Fathers. When they finally reach the mission at Zuni on November 24 they are so broken that they must rest until December 13 before continuing on to Santa Fe. They arrive at the capital on January 2, 1777. They have been gone 184 days and made a loop of 2,000 miles. And they have accomplished pretty close to nothing. Dominguez discovers he has been demoted from his post as head of the New Mexican missions and is reassigned to backwaters of northern Sonora. Escalante, a year after his return, is plagued by a kidney ailment and travels to Mexico City for treatment. He dies there in 1780.

What they leave is a shadow on the maps of the West, a dark spot that warns others. This darkness looms over southern Utah in general, and Paria and Escalante country in particular. Get Out If You Can. For a century other routes will try and slice across this continental belt—the route of Jedidiah Smith, the Mormon trail, the

Pony Express—and with one lonely exception all will steer far north of the canyons of the Colorado Plateau. Only the Hole In The Rock expedition will violate this cartographic taboo. About a century later, Powell's survey team will come upon a river spilling into the Colorado and casually assign it the name Escalante, though the good father probably never glimpsed the stream at all. The only settler to put down roots at the camp called *Salsipuedes* will be a fugitive from federal murder charges, a man seeking the most isolated locale in the United States. A man named John Doyle Lee.

There is something about this stone country that bewitches people. They spend their time both at war with it and yet under its spell. In 1889, the Denver, Colorado Canyon and Pacific Railroad Company is organized to find a route for tracks across this chaos of canyons. Robert B. Stanton is the engineer in charge of the survey. He never really recovers from the experience. For fourteen years he slaves away on a manuscript about the area, and when he finishes with it in 1920 it runs more than a thousand pages. He tries to see things as an engineer: "Every engineer—possibly there are exceptions—is possessed of two beings. With one he loves Nature for Nature's sake, loves it as God made it and gave it to us; with the other he follows Telford's definition of the profession of an engineer—'being the art of directing the great resources of power in Nature for the use and convenience of man.'" But the canyons break or seduce the engineer in Stanton. "I would not deprive anyone of the joy of seeing such a marvel—even from an aeroplane," he allows, "if he wishes, but, it seems to me, to really appreciate such a country one must see it, as one can only do, by wandering among its pinnacles and its gorges just as they were—untouched by the hand of man—years ago."[14]

THIS VISION COMES in the spring of 1860. The wind is bad, perhaps that is what triggers the vision. Or maybe it is memory, the guns, the women screaming, the yells, the clubbing, the children going down to their deaths. All those clumps of hair hanging on the bushes, sunbonnets fading in the sun. John Doyle Lee sees the way to be clean and he unflinchingly records what he sees. "I saw in a Night Vision," he begins, "the Process through which the Saints had to pass to become purified, which was as follows." There is a dungeon with foul air and Saints of all types are gathered there. Those possessed of the faith did well, but those who lacked the faith, "their flesh putrified & disolved & decomposed in ratio to the darkness that controled them." After this phase, the Saints move up out of the dungeon, and are laid side by side on screens for the second process. By this time some are so rotten that their bodies do not hold together when they are moved: "A more heart rending Scene I never beheld & would freely have given all I possessed to have been permitted to withdraw from the scene of horror, distress & Misery." But John Doyle Lee must work. He grabs a steel hook in each hand, swinging one into the jaws and the other into the ears, he pitches Saints up onto the platform. Here the violence of the steam hits them and they soften and melt and the flesh falls from their bones. He cannot take it, so he goes outside for some fresh air. But he returns to his duty. The third process is the shower bath, and the torrents of water wash the filth away. About half the people dissolve completely, maybe a quarter of the Saints come out half their former size. The rest stay beautiful and wear priestly attire. He ends with a simple question: "Shall I be able to stand the Test."[15]

THEY KEEP COMING into the country and never quite know why. Everett Ruess wanders the canyons in the early 1930s with a burro and his art materials. He drops down into the Escalante and camps in Davis Gulch, where he scratches the word NEMO on the rock—Jules Verne's work is one of his favorites. He is twenty years old in November, 1934, and he disappears forever. People still look for him, but I do not. He is where he was headed.

I meet another one, and he, too, has been drawn to his Escalante for reasons he cannot clearly state. His story is very simple. He is living in Germany in the early seventies and is very interested in early American civilizations. He reads a great deal. He also runs a drug laboratory. The lab hides behind a hedge in a small cottage, and one day, as he approaches his little illegal factory, the hedge explodes into flame. He backs off and the fire dies. Three times he tries this, and each time the same thing happens. That night he sees strange things in the sky that he believes are messages. The next day he is on a plane to Utah. He becomes a Mormon, moves into a teepee, and lives on a rock. He eats only vegetables now and travels around the country with a small figure of Elvis Presley and another one of Marilyn Monroe. He poses them and takes photographs in various settings. But he always returns to the Escalante and his tent. I believe he is working on a script about the Shroud of Turin—but one with a twist he tells me. He laughs a great deal now.

She was a dental hygienist back East. I forget how it began with her: a river trip? a backpack? Anyway that was decades ago and now she guides people into the back country. She horse packs, runs rivers, takes clients along on round-ups, trout-fishing expeditions, what have you. She lives in a hamlet and in her spare time trains her horses. She probes every nook and cranny of the Escalante and the other canyons and is disturbed by the fragility of this world of rock. One year at a hunting camp she spills some oats she is feeding her pack train. When she returns the next year, the whole campsite has metamorphosed into an oat field. It can change like that, with one careless error.

Now, of course, the plateau is changing everywhere, with the explosion of second homes for the rich and the bigfoot prints of industrial tourism. We are drinking a beer, she is talking and laughing. She is living off this new wave of people who want to hike and fish and raft and ride mountain bikes and pretend to be cowboys. What can she say? But it seems remote, this world she works in, from the thing that brought her here. A thing she cannot seem to express. But then no one ever really has.

They'll tell you of a canyon, and to get there you must tackle a few bad pitches and do a short rappel. You will find a grotto with ferns and moss and blood-red rock and water trickling down the cliff face and a pool at the base. You lean over and see your face reflected in the water, though you cannot really recognize it. They'll tell you of riding horses in the Henry Mountains, and you are at ten thousand feet and the air is cold, the forest green and awash with life, and then you will ride out into a clearing and will be able to see, and the buffalo will stare back at you with their black eyes. Or it is the trout plucked from the waters, the frying pan near at hand, that sizzle and the succulent flesh followed by a glass of wine that magically appears. At night the sky will be full of huge stars that hang like lanterns. They'll tell you of trips where the last day out you are exposed on the rock, and then when you make the rim there are still miles of hot sand and no shade. They'll tell of the river trips gone bad, the raft capsizing, and you are pinned underneath, the brief panic, the fight to the surface, and then you bob like a cork in your life jacket before being swept into the boulders and noise of the rapids. They tell you of various hogbacks, corkscrews, combs, mazes, dives, rapids, waterfalls, slickrocks, and pot holes. But mostly they'll refuse to tell you much at all.

It is not simply that words cannot express what they feel, it is more a hesitation about saying such things. It is almost a taboo—as if the straitjacket of language will kill the thing and make it all go away. They are right. There are many fine places in this nation but few in which anyone has attempted to found Zion. And all the failures of the canyon country are not really bitter but comforting. No one can possess this rock. They can pretend. They can buy and sell and build and throw out redwood decks like blankets. But they cannot possess it. No one ever has. It has remained stubbornly empty, a thing to go around, to avoid. And it will tame us also, or, if we persist, break us. But it will have its way. And so we will keep going back, the truck will get stuck in mud, the waterhole will be dry, the night will bring a storm that freezes our red sunburned hide, and the dawn . . . well, it will sound like a harp or a flute or pure silence.

As I said, they'll tell you a lot of things. But mostly they'll refuse to tell you much at all. You'll come, and if you want it, it will come to you. And then you will have things to tell and you will tell them over a dinner or a drink. We'll all be the same then, talking away. But sharing this silence about all those other matters.

Now, there's this highway that cuts through flats of sand and little bitty scrub and along that highway you see honest-to-God, official yellow triangular signs that warn Watch For Eagles On Roadway. Well, where else are you ever going to . . .

In the Heart of Stone

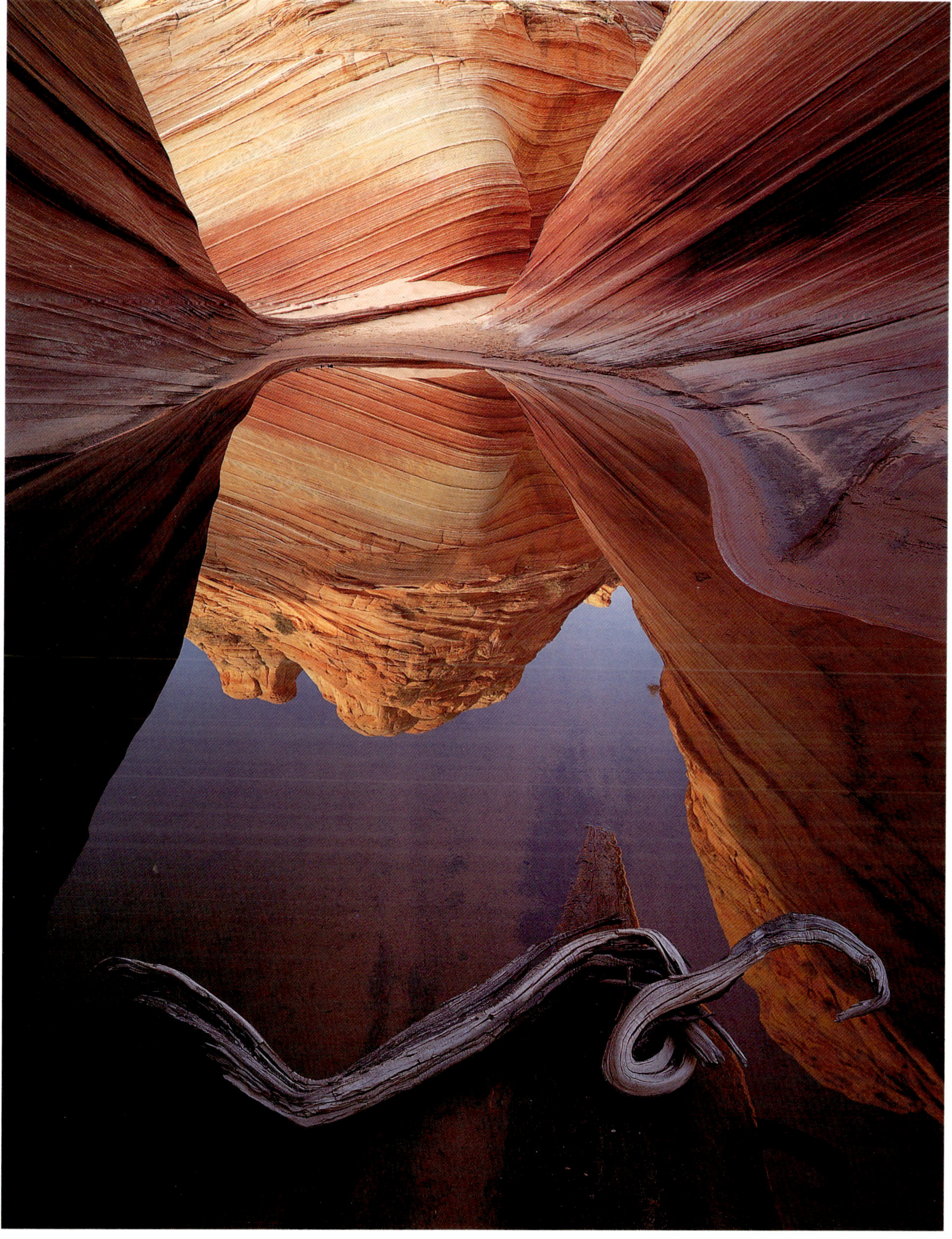

Petrified sand dunes border a well-watered pool. (Paria Canyon, Vermilion Cliffs Wilderness, BLM, Arizona)

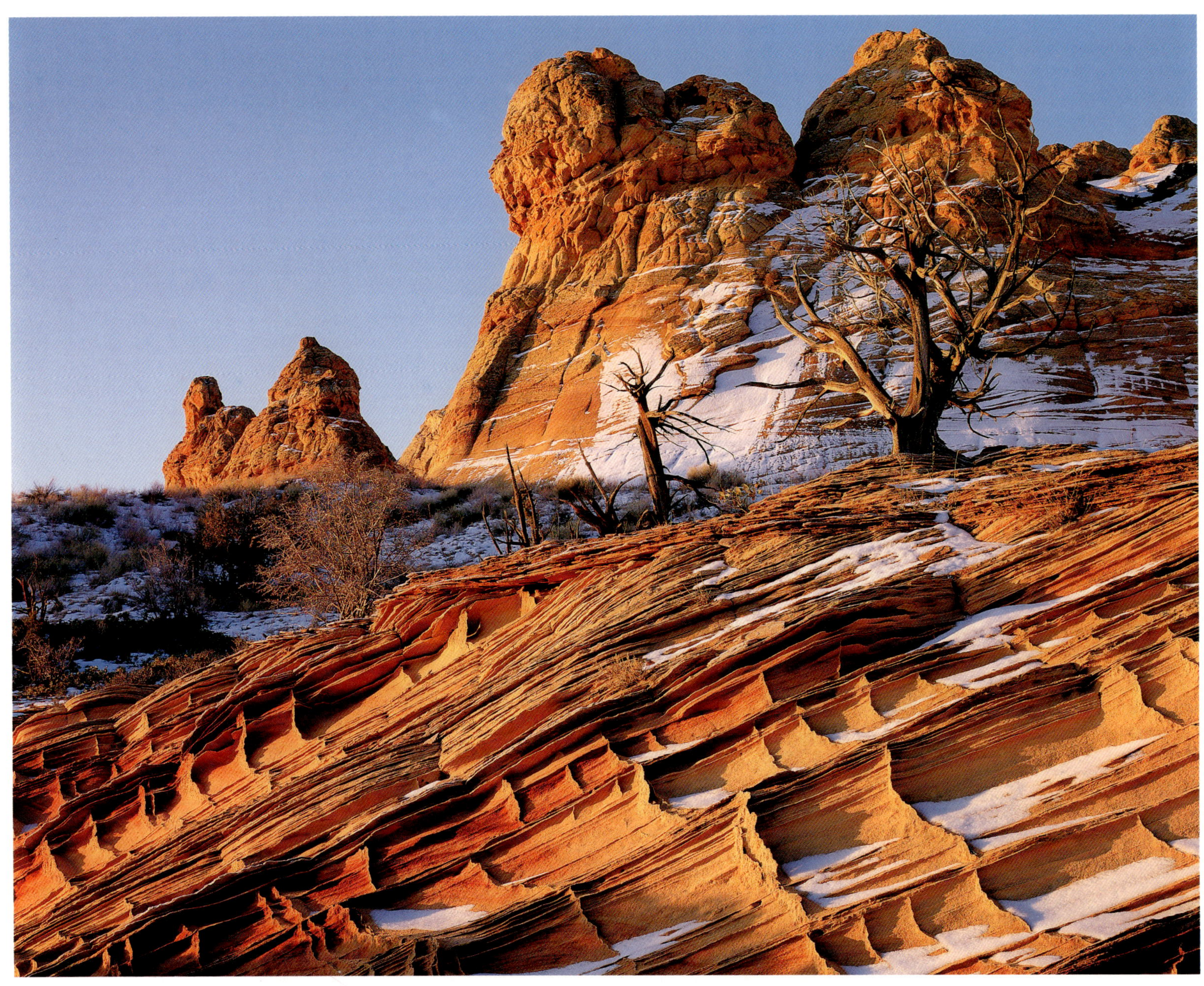

Spring snow flecks the crosshatched sandstone strata of petrified sand dunes at dawn. (Paria Canyon, Vermilion Cliffs Wilderness, BLM, Arizona)

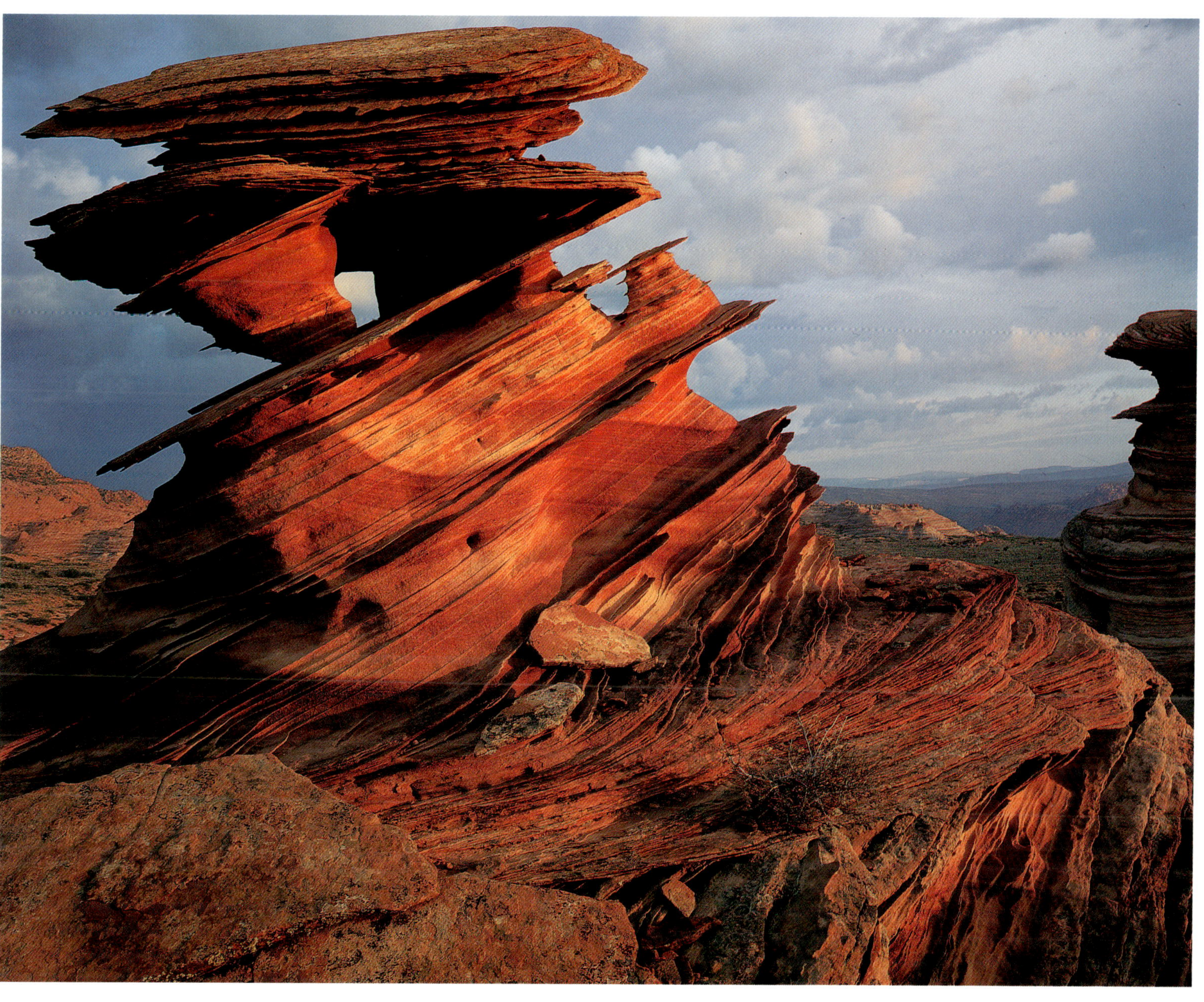

Morning light picks out the delicate bands of eroded sandstone in petrified sand dunes.

(Paria Canyon, Vermilion Cliffs Wilderness, BLM, Arizona)

Roots of a dead juniper point skyward among petrified sand dunes. (Paria Canyon, Vermilion Cliffs Wilderness, BLM, Arizona)

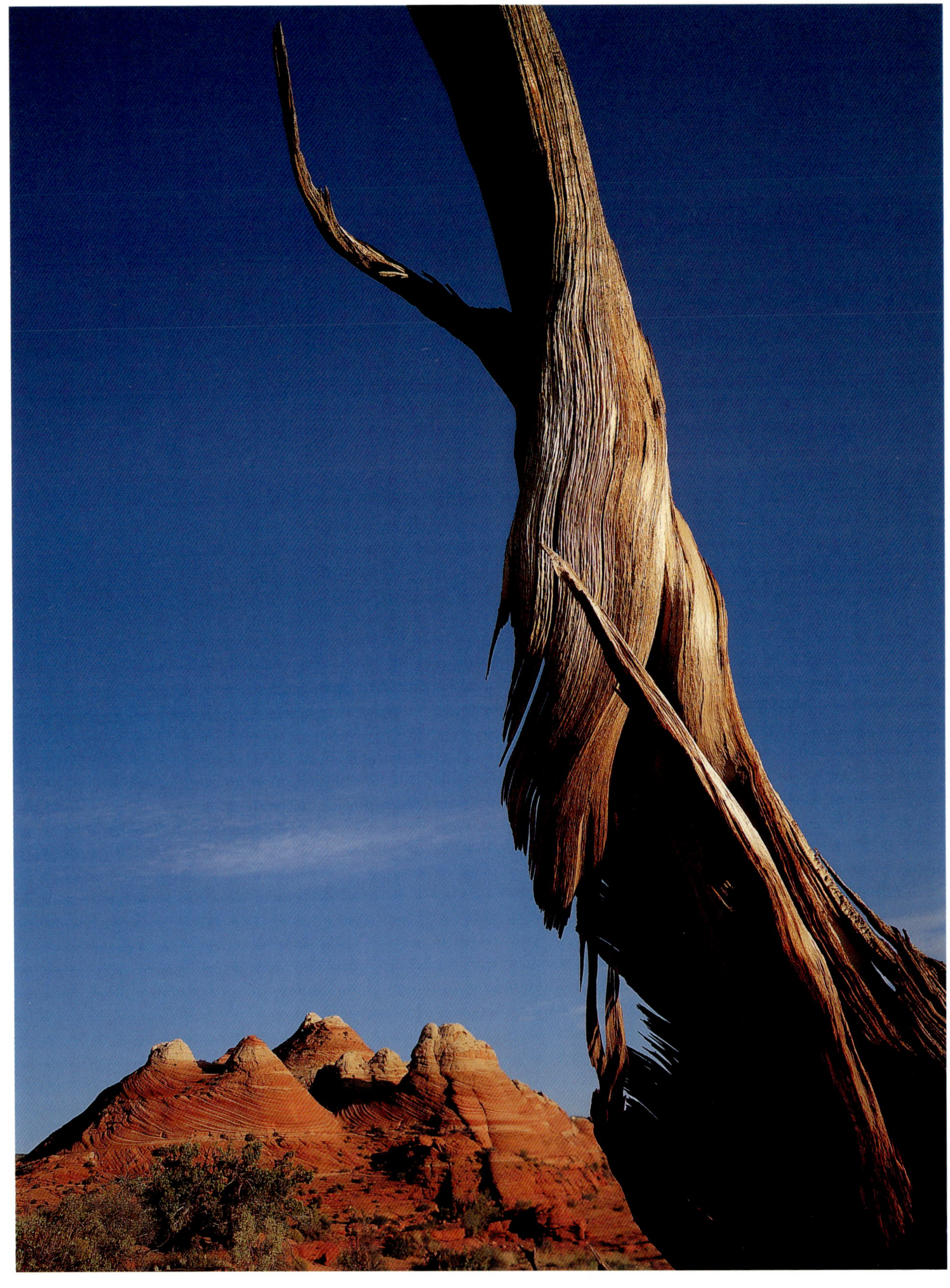

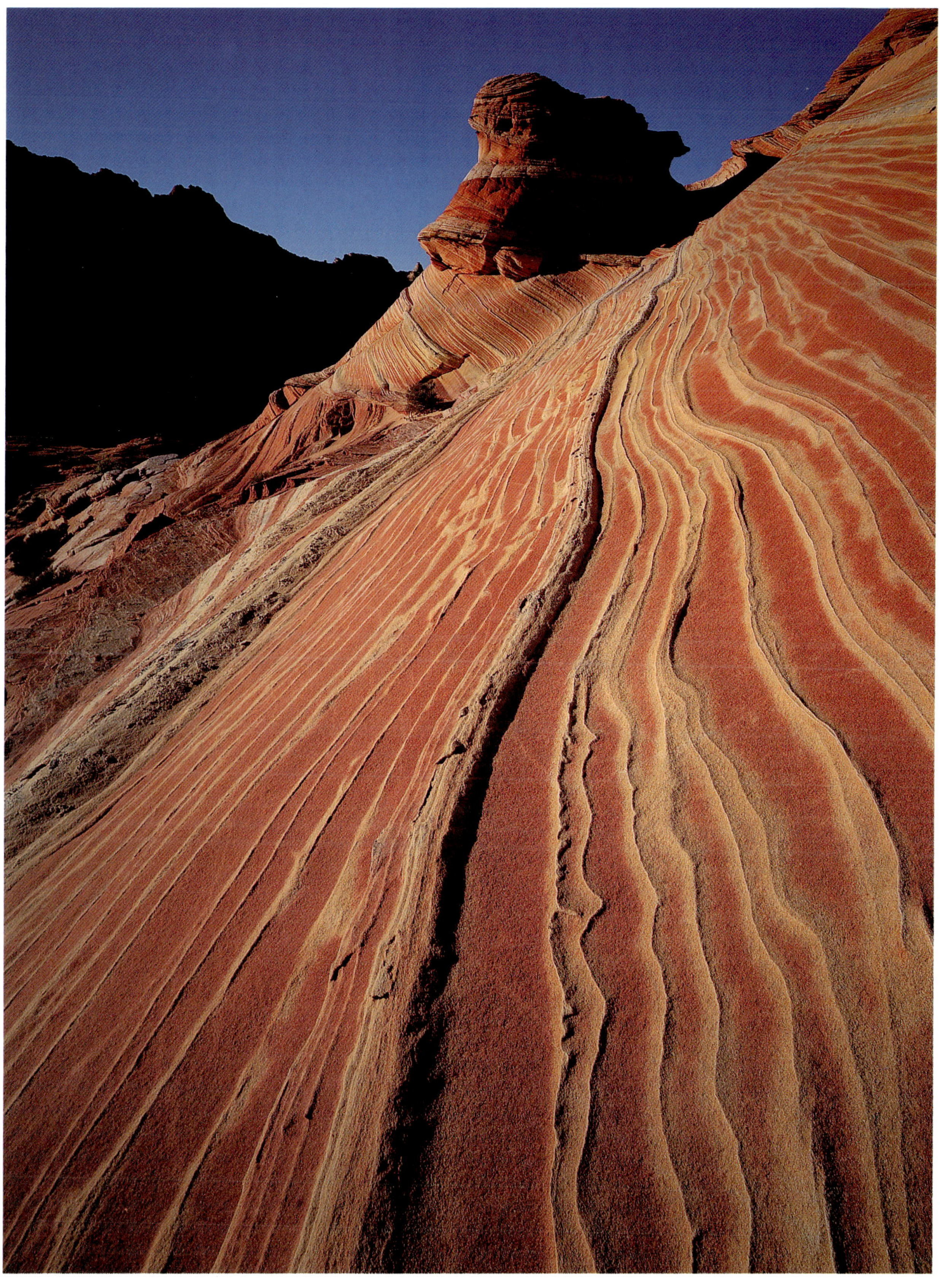

Bands of red and yellow mark the strata of petrified sand dunes at sunset. (Paria Canyon, Vermilion Cliffs Wilderness, BLM, Arizona)

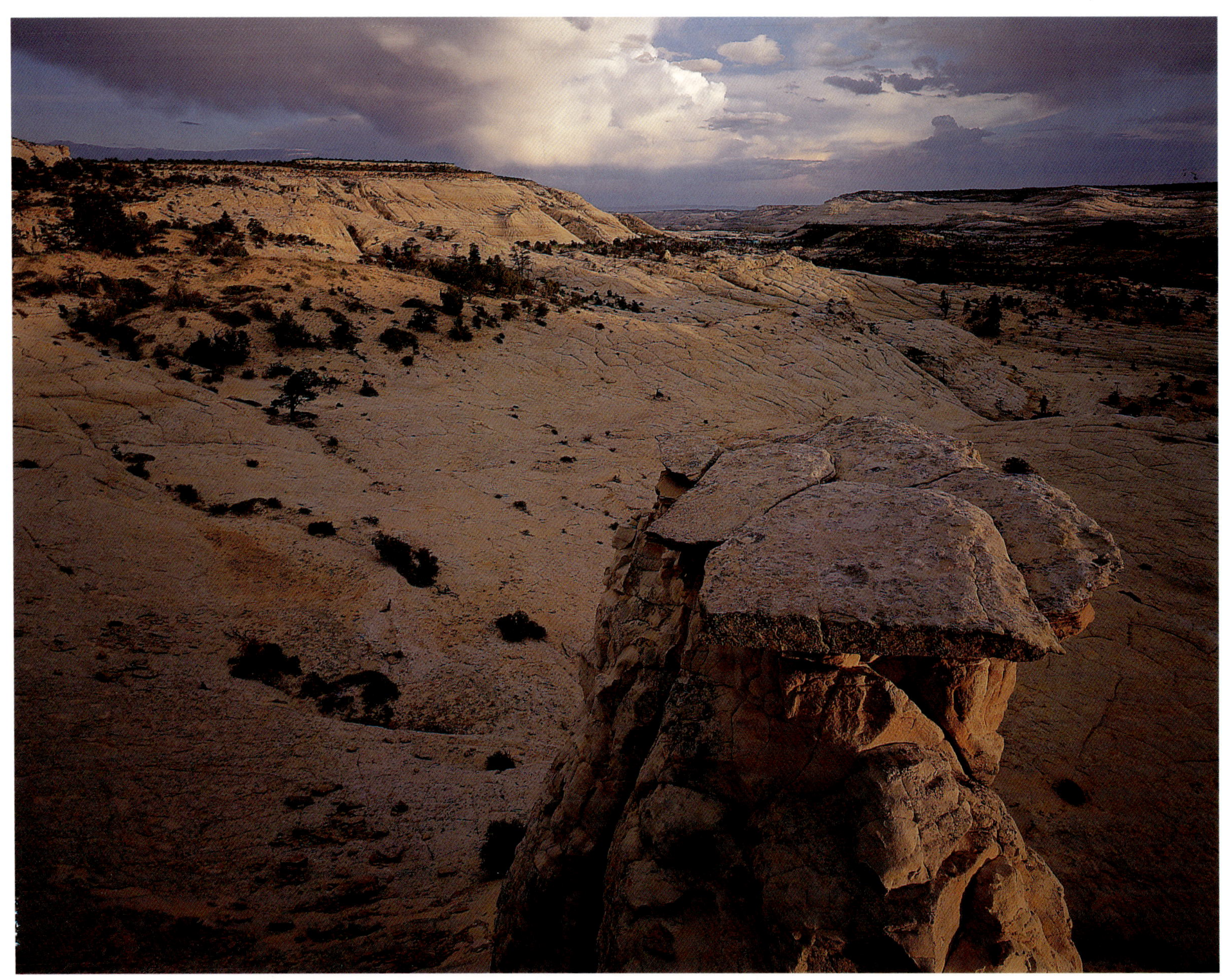

Storm clouds at dusk loom above a balanced rock at the edge of the slickrock wilderness above Sand Creek. (Escalante Resource Area, BLM wilderness study area, Utah)

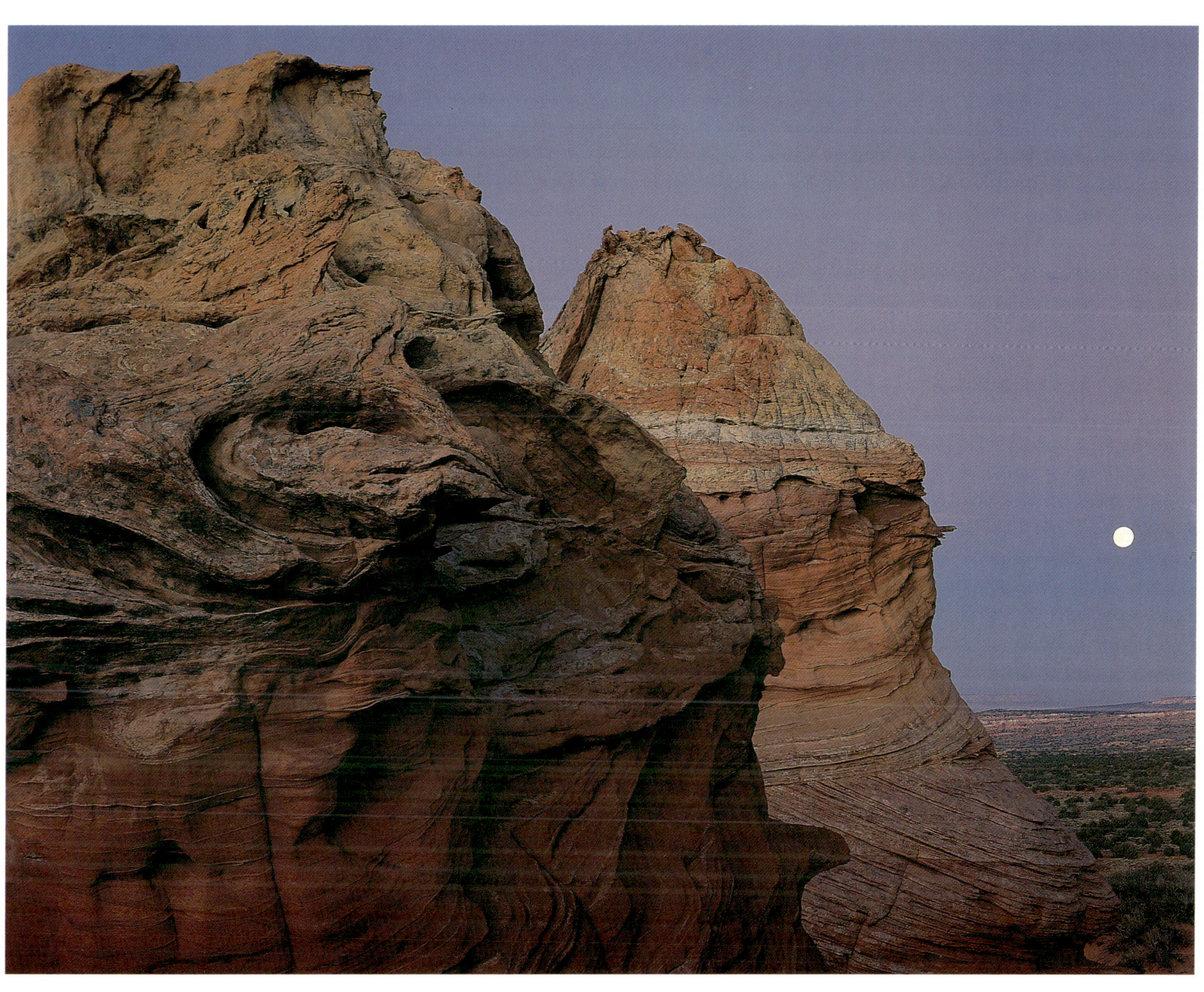

A full moon rises above the petrified sand dunes. (Paria Canyon, Vermilion Cliffs Wilderness, BLM, Arizona)

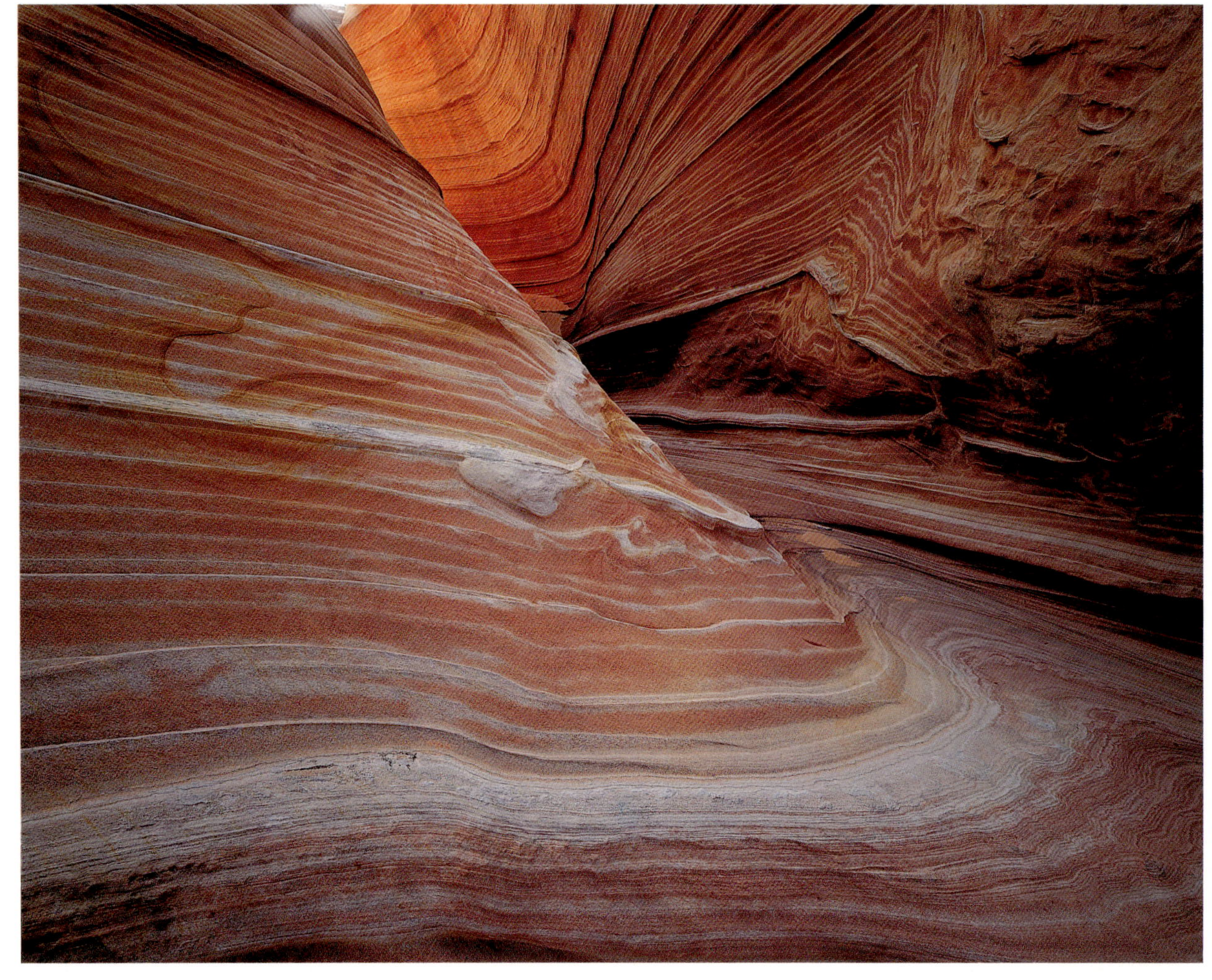

Petrified sand dunes with yellow, gray, and red stratified formation bands. (Paria Canyon, Vermilion Cliffs Wilderness, BLM, Arizona)

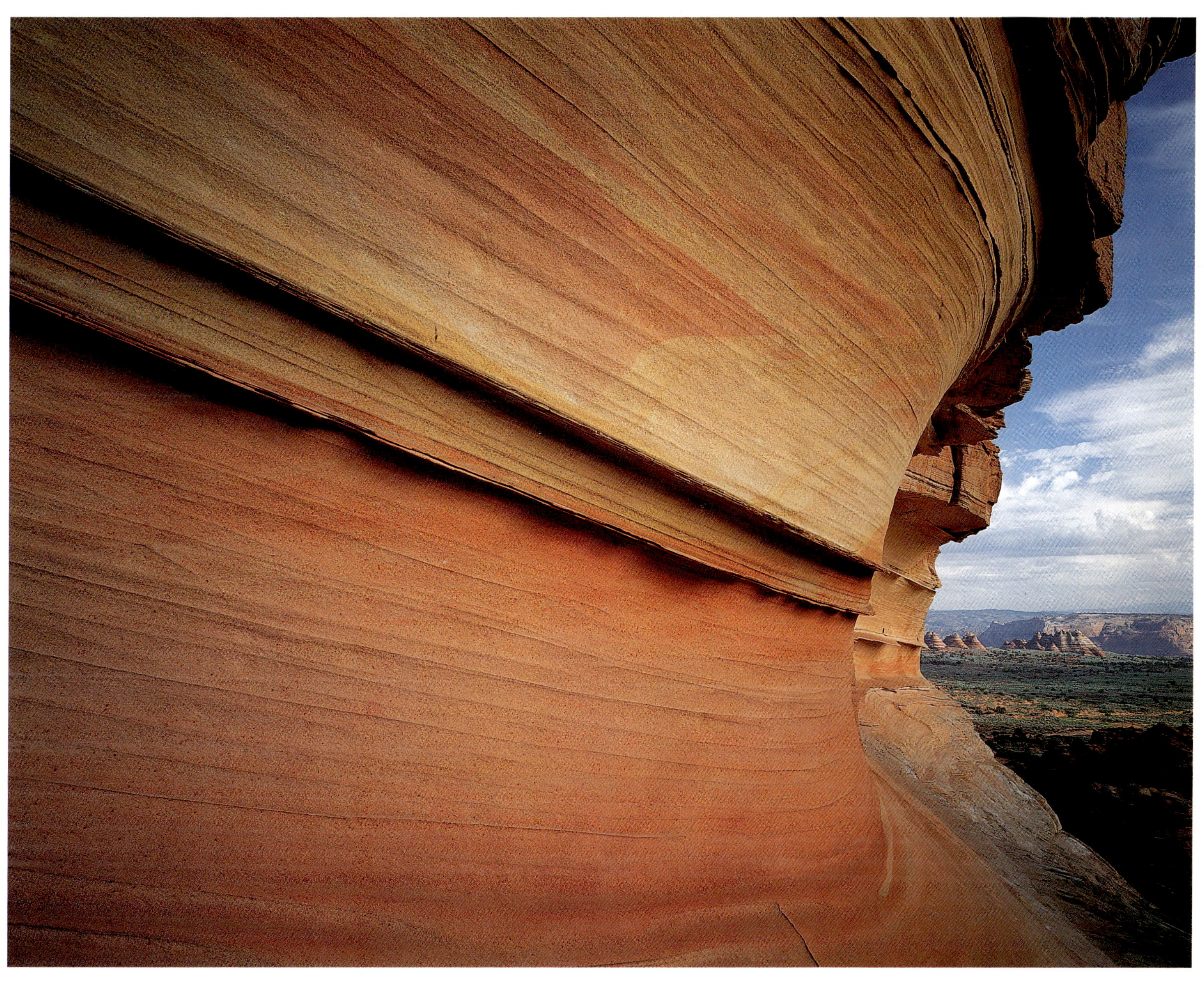

Morning sun lights the petrified sand dunes. (Paria Canyon, Vermilion Cliffs Wilderness, BLM, Arizona)

Desert varnish stripes and Douglas fir (Pseudotsuga menziesii) *in Long Canyon. (Escalante Resource Area, BLM, Utah)*

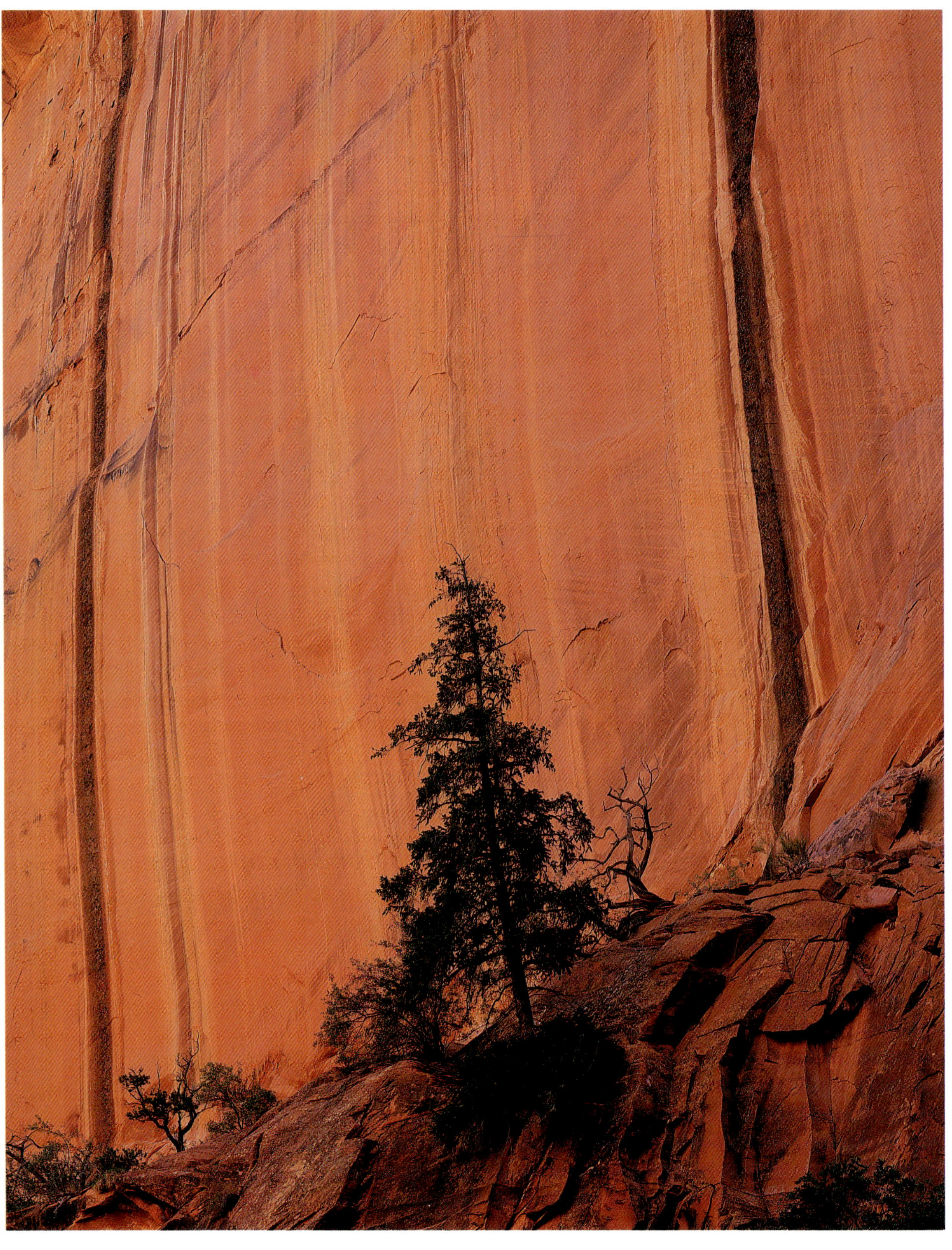

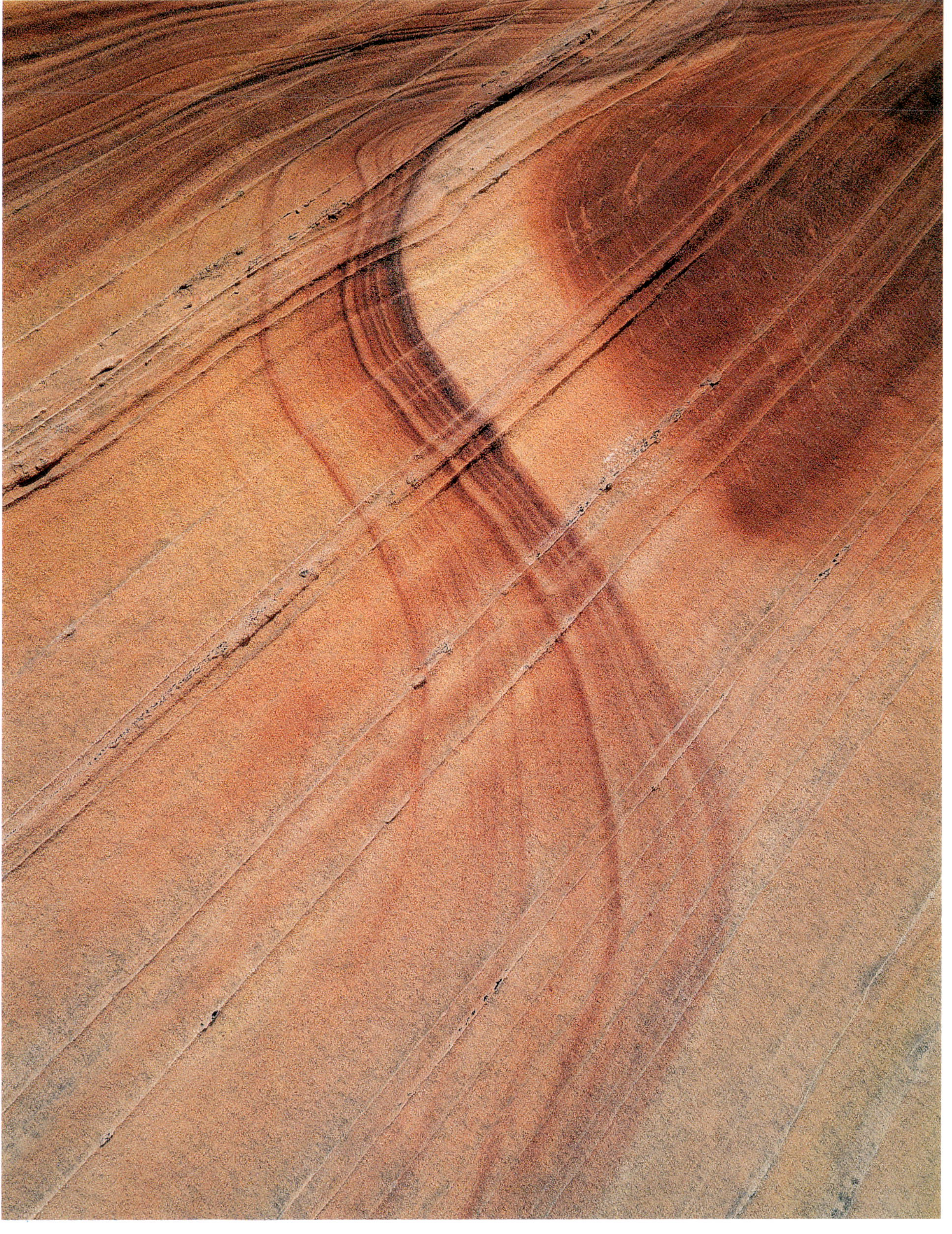

Swirl patterns from mineral deposits mark the eroded sandstone of the petrified sand dune formations. (Paria Canyon, Vermilion Cliffs Wilderness, BLM, Arizona)

Wind-sculpted sandstone and twisted roots among the petrified sand dune formation. (Paria Canyon, Vermilion Cliffs Wilderness, BLM, Arizona)

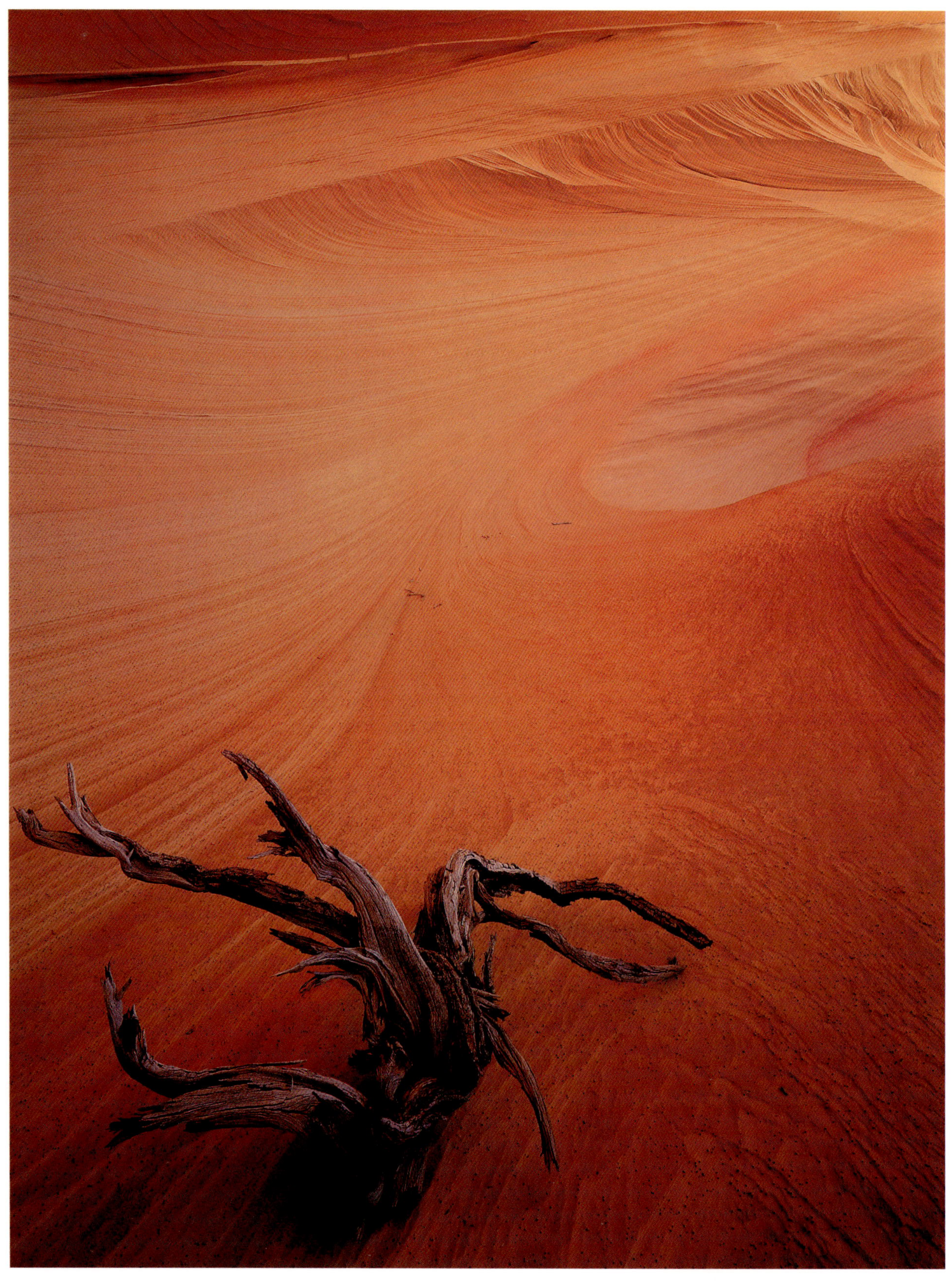

Windblown pink sand drifts among water-polished gray sandstone. (Escalante Resource Area, Wilderness Study Area, BLM, Utah)

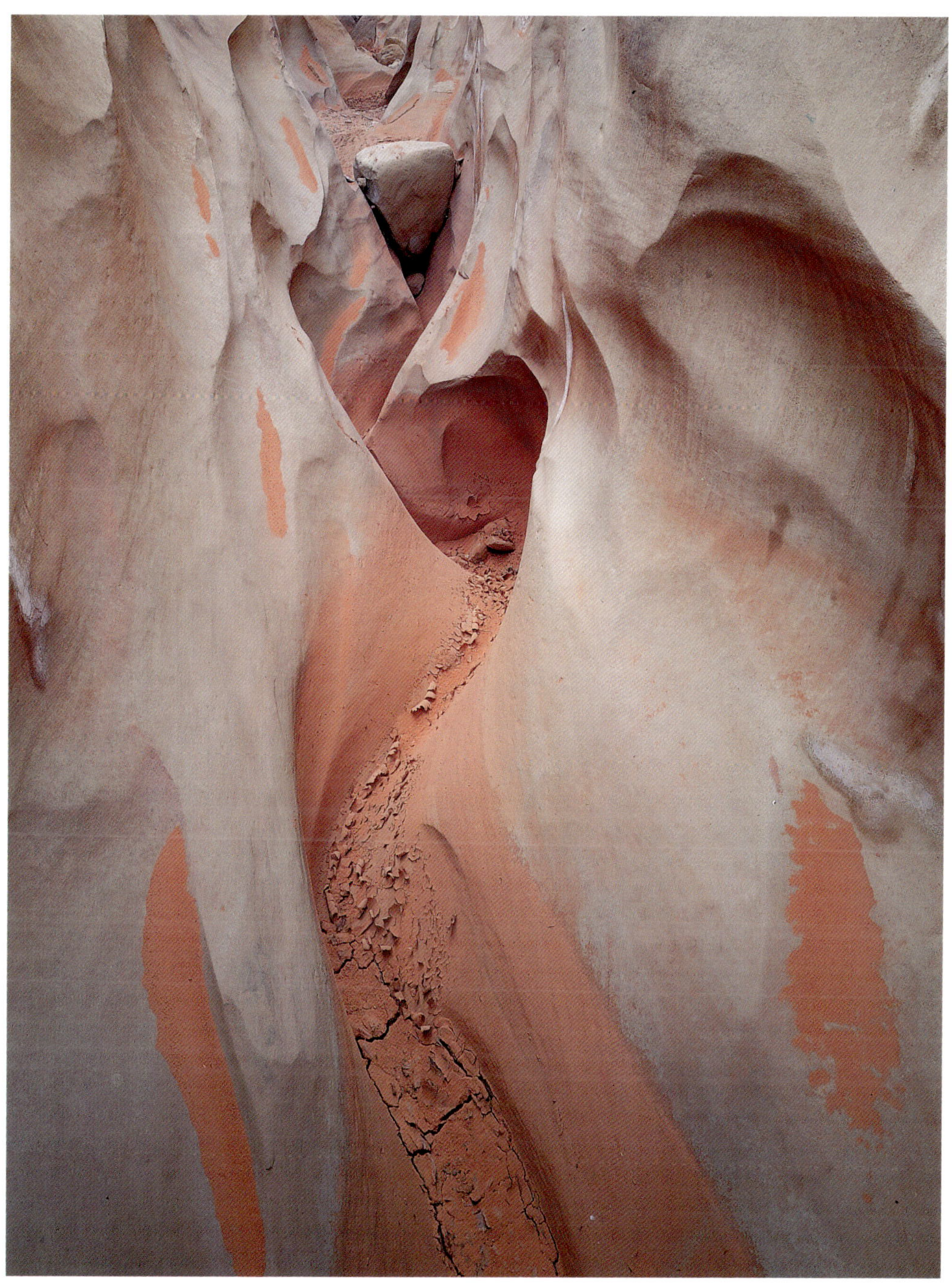

Autumn-colored boxelder (Acer negundo) *at the Escalante Natural Bridge. (Escalante Resource Area, Wilderness Study Area, BLM, Utah)*

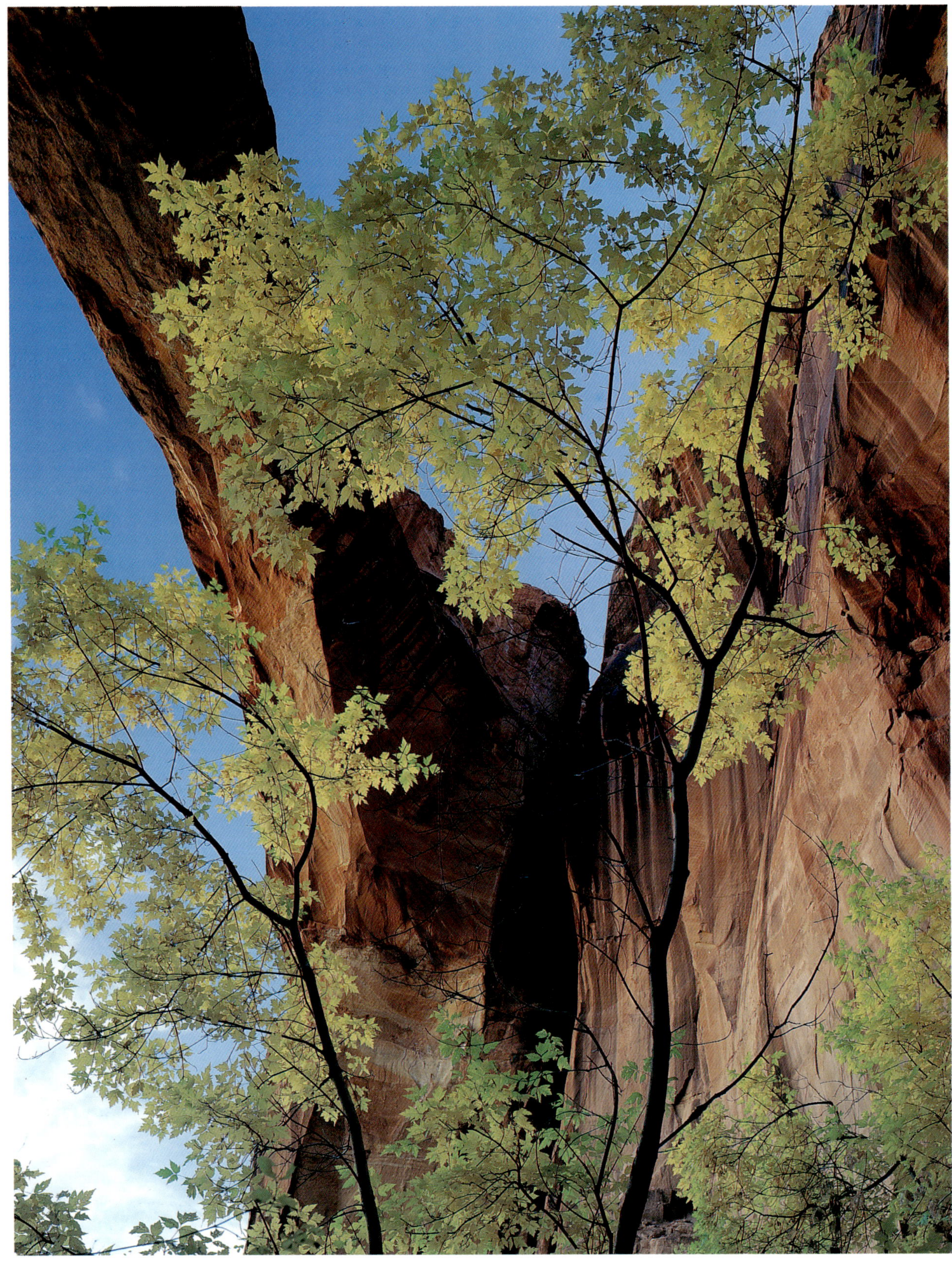

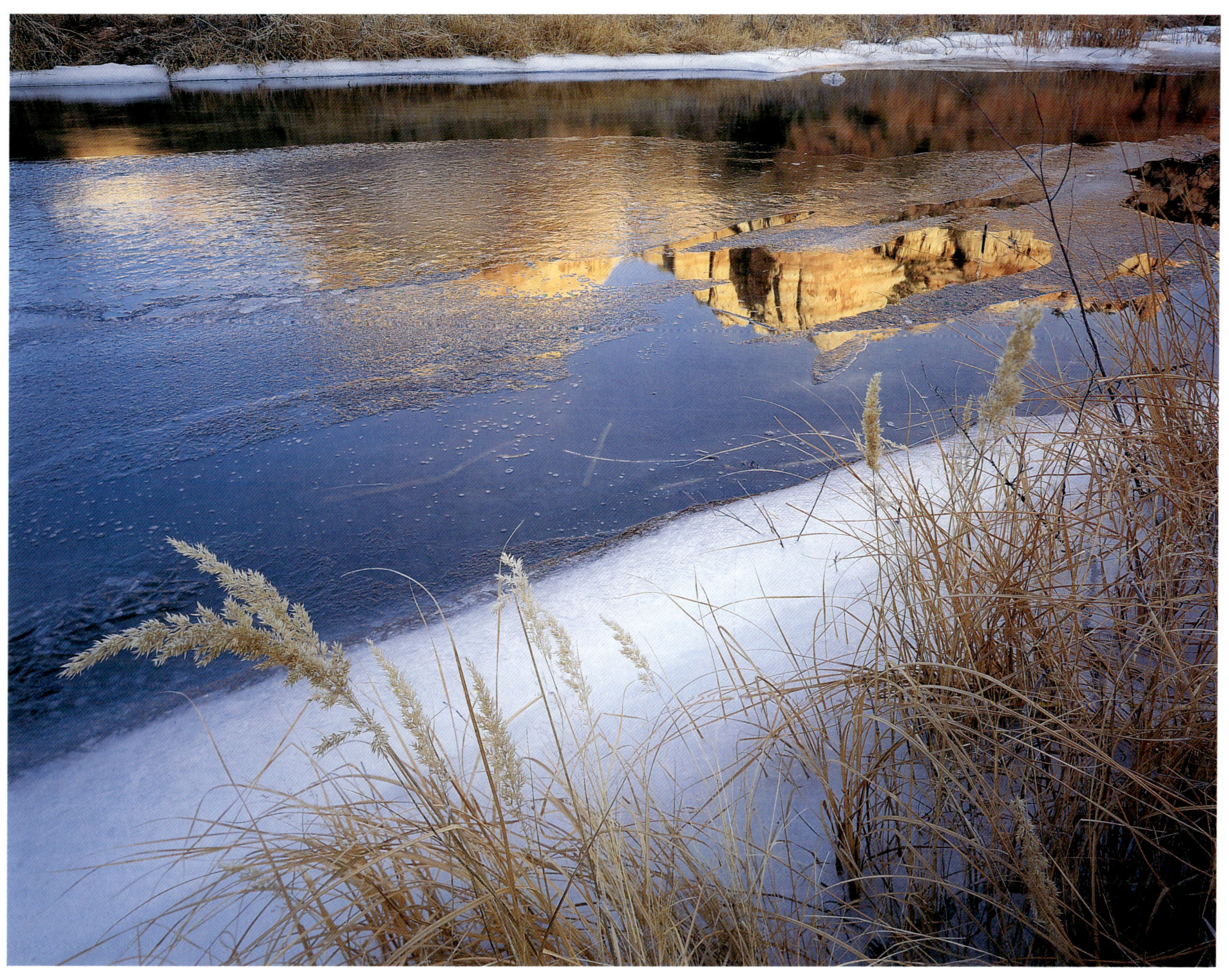

Grass-lined banks frame cliff wall reflections in ice-covered beaver ponds at Calf Creek.

(Escalante Resource Area, Wilderness Study Area, BLM, Utah)

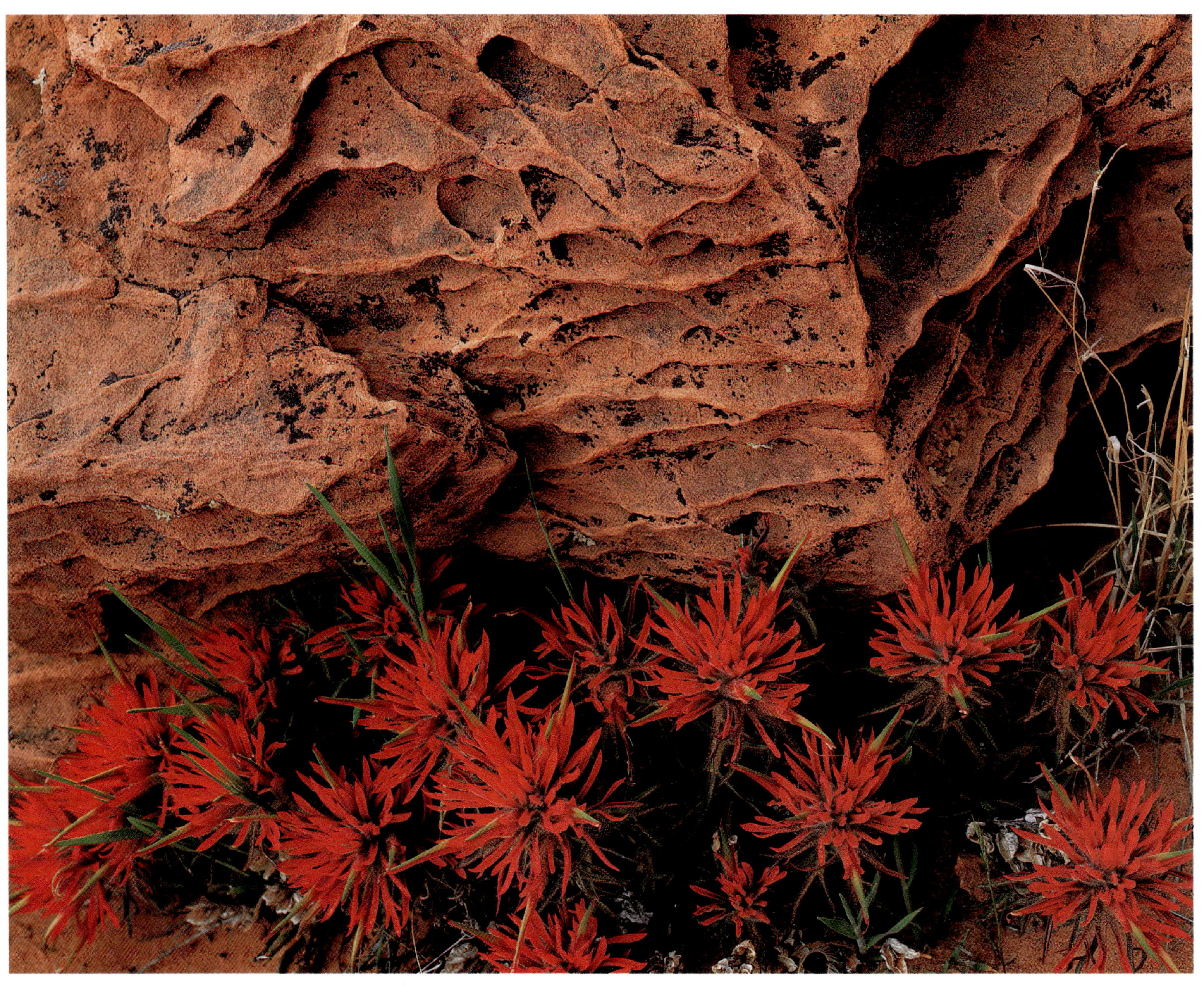

Indian paintbrush (Castilleja sp.) *seen against eroded sandstone. (Paria Canyon, Vermilion Cliffs Wilderness, BLM, Utah)*

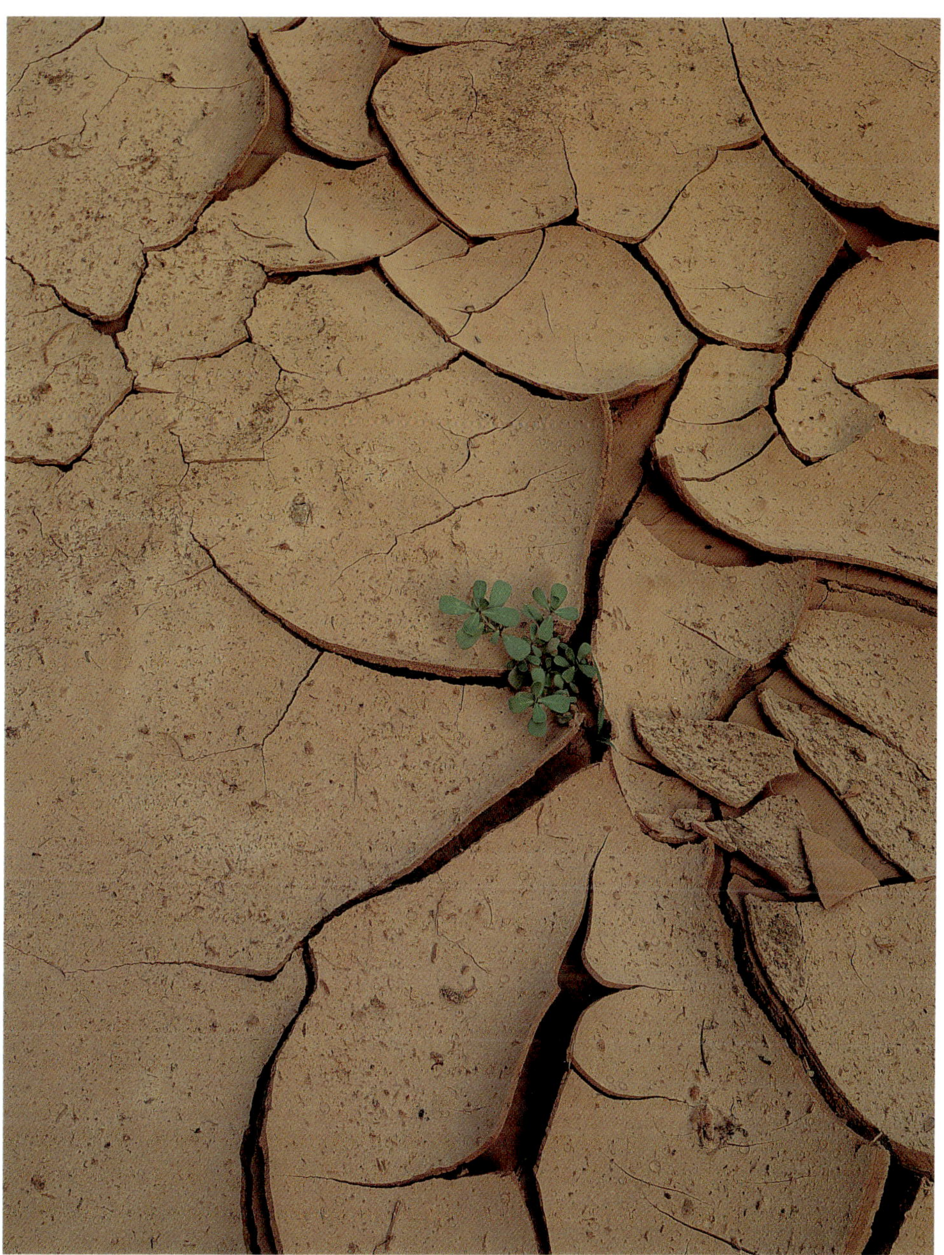

Purslane (Portulaca oleracea) *ekes out a living in the cracked clay of the streambed of Buckskin Gulch in Paria Canyon. (BLM, Utah)*

Flowering Indian paintbrush (Castilleja sp.) *grows amid a stream's boulders. (Death Hollow, Escalante Resource Area, Wilderness Study Area, BLM, Utah)*

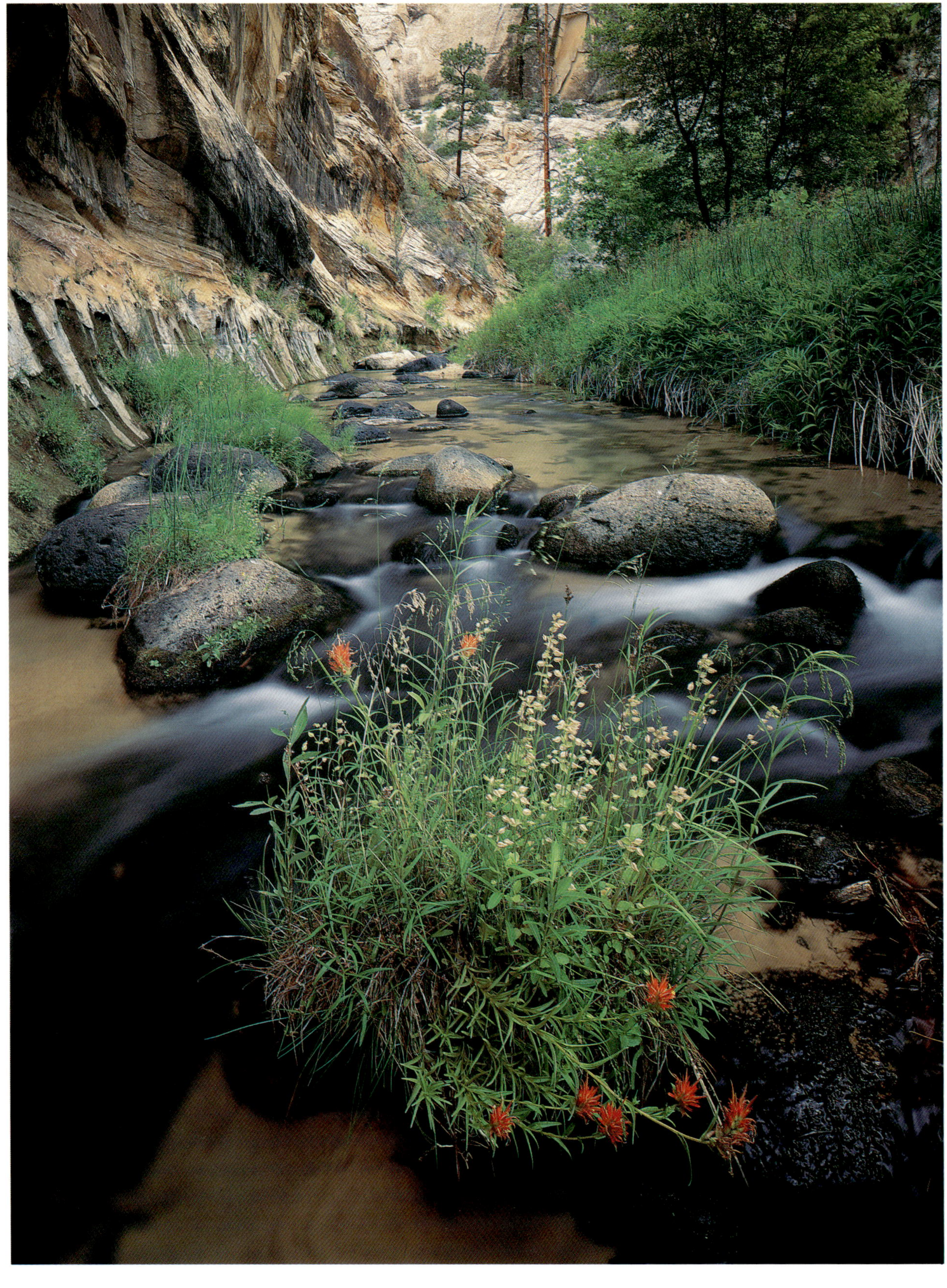

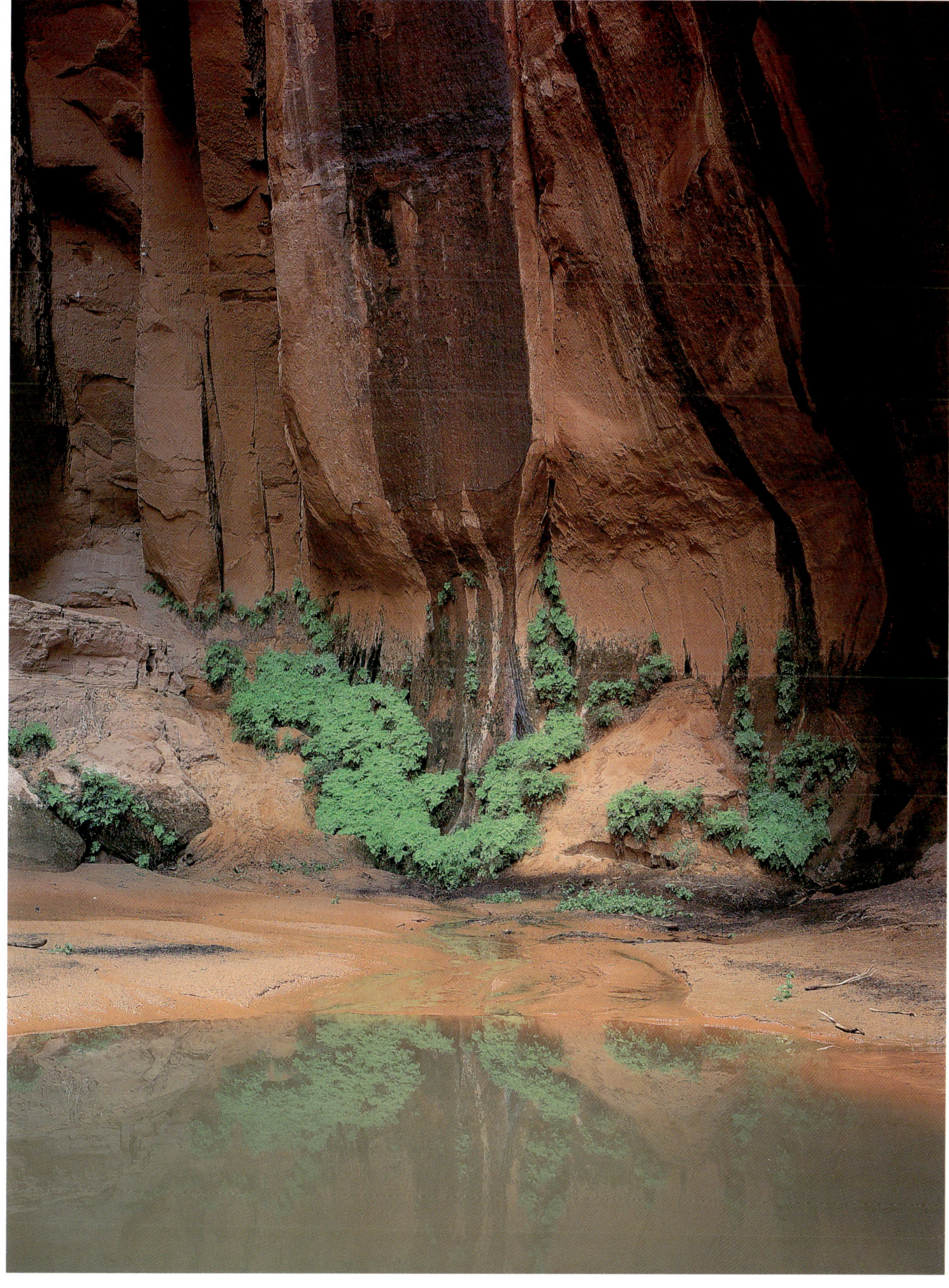

Lined with maidenhair ferns (Adiantum capillus-veneris), *the "Hole" is a side canyon of Paria Canyon. (Paria Canyon, Vermilion Cliffs Wilderness, BLM, Arizona)*

THE HOLE IN THE ROCK expedition begins with jabs and feints on endless rock. An early party drops down the Paria, crosses where Lee had once run his ferry service, scrambles up the stone torture of Lee's Backbone, and then wanders through the Navajo country until finally bursting out of the high desert at Montezuma, Utah. They are half starved and spooked by hostile Native Americans. They have spent weeks digging wells by hand at dry camps, and barely slaking the thirst of their livestock or themselves. They promptly set up a Sunday school, start laying out fields, and celebrate the Fourth of July. Elizabeth Harriman whips out an American flag by snatching some blue from her little girl's dress and getting the red from Zechariah B. Decker's long johns.

But the route they have taken does not tempt them on the return trip of some of their party. Instead they go far north, then west, and finally south again. It is roughly two hundred miles or so from Escalante, Utah, to Montezuma. But the circuit sketched out by this exploratory group is almost a thousand miles. And it reveals one telling fact in its huge arc—it avoids every square foot of the Escalante country. Clearly, this will not do for the main party of people following. It is too long, and the section through the Navajo ground too dry and dangerous. So a shortcut must be found.

Men are dispatched to explore the jagged cliffs framing the Colorado River. One Saint sees a deer, and while pursuing the animal watches it jump down into a slot and vanish from view. He goes over and finds a deep slit leading down to the river, a place that will become known as the Hole In The Rock. There are other probes, but this almost clifflike face plunging down to the river is what will launch the caravan of settlers and their stock toward Montezuma and their mission. They will find a way to take a wagon train where a deer has bounded down out of view. It is seen as a wonderful shortcut. This is a necessity: the people must arrive at their destination in time to plant spring crops, get in a harvest, and prepare to winter over in their new world. Besides, they are Americans and there is no time to waste. And besides being Americans they are Saints, who feel the final days are never far from their lives. They have lived in a world of lightning change. A brief generation before they had walked across the West to reach Salt Lake, and now the nation is welded together by rails and driven by the exacting schedules of locomotives. Their impulse to cut corners, drive hard, and watch the clock will never be erased from our national psyche.

I AM WALKING THROUGH the desert near the route taken by the Hole In The Rock expedition. The sun is beginning to flatten out the land as the glare of midday comes on. A rooster tail of dust approaches me from a two-track desert jeep road. Suddenly, a station wagon brakes to a stop and four men leap out togged in Lycra britches and shod in the comic footgear of mountain bikers. They are waving maps at me, asking where they can find the trail head. They do not smile. This is serious business. I point out a route, and they roar off with four bicycles lashed to the roof of the car. No time to waste in this leisure business.

They are the descendants of the Saints who decided to take a short cut they had never really explored in order to save that precious time. We are all their descendants, as our digital clocks keep reminding us. So the Saints pull out. They do not know that their short cut, the dream of a route through the Hole In The Rock will take a little longer than they had hoped. In fact, it will take them as long simply to reach the Hole In The Rock as the entire journey would have required if they had taken the impossibly long arc around the Escalante country that first exploring party had favored on its return route.

But they cannot wait and the groups of wagons converging from all over Utah on the hamlet of Escalante start streaming south in October of 1879. Their rendezvous point is Forty Mile Spring, a bit more than halfway to the Hole In The Rock. By mid November wagons are gathering there. Four weeks have passed of a journey projected to last only six weeks in all. There are three fiddlers in the company, and the young couples gather in the evenings at a nearby sandstone amphitheater they dub Dance Hall Rock. Snow is beginning to fall, there is no time to waste.

And then a report comes back from the cleft they have called Hole In The Rock. The survey of this crack concludes, "If every rag or other property owned by the people of this Territory were sold for cash, it would not pay for the making of a burro trail across the river." But it is too late. They are committed. And they have no time, no time at all.

JOHN DOYLE LEE is ill, but he can communicate through the birds. He is drinking turpentine to ease his pain as he explores the deserts of northern Arizona. He has sent a message to one of his wives for medicine, but this will take five days, assuming that the medicine ever comes at all. He is too weak to make a fire or walk to get water. He lies down to meditate and then, "I saw Rachel A. [one of his wives] come & bring me a fish and I had the offer of Riding in a beautiful new carriage drawn by a Span of dark bay Mules but refused as I choose rather to go in search of My sheep that I Just heard from." He looks over and sees a flour milling going full blast and the miller is a dead man he knew. "I awoke," he jots in his diary, "& was Satisfied that the mission would go on & I would have to go with & explore the country. . . ." He prays to God to send a message to his family and makes up verses:

> Fly my Sweet Bird to my House in the Cove
> & whisper this message to my loved ones Home
> Tell her to come Quickly, My own Bosom
>   Friend,
> For I am alone in deep anguish and Pain.

"Just before penning the same," he notes with wonder, "a Bird came & lit within a few inches of my head & appearingly tried to talk. I coaxed him a little bit & then bid it go."[16]

Of course, he can be a different kind of man than this bird talker, or perhaps, simply a younger man. In 1848, almost a decade before his actions at Mountain Meadows, he supervises a hunt suggested by Brigham Young to rid the area of wolves, wild cats, skunks, minks, bears, mountain lions, coyotes, eagles, crows, ravens, and hawks. Two teams of about one hundred men each set out to slaughter with a point system agreed upon to determine the champion. Between fourteen and fifteen thousand animals are killed with Lee himself accounting for 2,043 "skelps."[17]

Perhaps it is all a consequence of truly coming into the country, of realizing the immensity of the rock and canyon land and hard desert that makes him more aware of the things around him. On August 1, 1873, he is thinking again of that bird that landed by his head a week before. The event bewitches him so much that he rereads his earlier journal entry. He realizes now that as he looked at that bird the thought crossed his mind that he could send a message to his family through its auspices, a message "to let My Family know My lonely & forlorn condition."

Later he learns from his wife Rachel that, "on the Same day about noon a Bird came to the window. Finding that it could not get in there, came in at the Door, lit on the window Sill by them, appearantly tired & hungry, picked up some Bread crums that lay there. Rachel Said that it was a messenger from Me, & fed it more Bread. It Eat without fear, Sit on their hands & tried to talk. It was a Singuelar Bird from any that had even Saw before. When it had refreshed it Self, Rachel Set it on her hand in the Door. It turned to wards her & bowed twice & took its leve direct towards My retreat. They all looked after it till it was lost in the distance." Rachel tells the children that her husband is in distress and makes ready to go and find him. "Thus," Lee realizes, "was my Prayers answered, & that, too, in a Maraculous Manner."[18]

Of course, no one today can fully accept such a story. Just as no one who has been into the stone and sand of the country would be quick to question it. He is in these moments all of us, taking, slaughtering, diminishing and yet at the same yearning, praying, hoping for some bridge to this world that surrounds us. Ah, we no longer slaughter the predators wholesale and think we are doing a good deed. But we no longer listen to the birds, either. We now know more names, but they experienced more places. We spend our time chasing their strange shadows, these people who came into the country and truly tried to stick; who lived the ground and, of course, were defeated by it, though they were tougher than we are. And a people of great soulfulness.

I LIKE TO LIE OUT at night. I was born to hate tents.

The stars wheel overhead and the heat leaves the land soon after sunset and the chill comes on. Everything is rock and it wants to suck the heat out of my body. I am stretched out on solid stone by the fabled Hole In The Rock. Bats dart over my head, and their swooshing past makes me feel less alone. The stars outnumber me by a billion or two and are no comfort. But the bats are similar bloods and there is a bond. In a while the moon will rise and ruin the sky with its milky glow. My time is now while the light night breezes purr, the bats flitter and feed on insects, the burn of this day's sun still flames off my reddened face. As I watch the night flights I feel I am being woven into a cat's cradle of energy. Perhaps, later I will beckon one of the flying beasts and send a message.

Whisper this message to my loved ones at home. Tell her to come quickly.

THE HORSES STRAIN crossing the gulches and the gulches get progressively worse as the Hole In The Rock expedition moves from Forty Mile Spring and the pleasures of Dance Hall Rock toward the stone lips of the Colorado Gorge. The fine red sand is a curse, and when the wind blows it scrapes against members of the party like vicious snow. At night there is little or no wood for a fire, and they make do with shadscale, a small bush. A big pile of the plant will make a good fire for maybe half an hour. They are now eight weeks into a six-week trip and food is scarce. There is almost no game in this dry country and the Saints resist slaughtering the cattle upon which rest all their hopes of future herds. But eventually the killing begins. By December 11, 1879, Platte D. Lyman writes dryly, "Killed our beef in the morning and loaned most of it to the camp." Flour has almost vanished and by the time they reach Fifty Mile camp they have jerry-rigged a coffee grinder and are devouring the seed stock they will need for planting. The mill runs all day and sometimes the grinding continues until midnight. Elizabeth Decker notes, "We have just sent our last five dollars to Escalante to get some pork and molasses."[19] The livestock also has little to eat. Even mules begin to give out from weakness. They decide to drive the horse herd down to Jackass Bench to keep them alive. But nine fall off the trail and die. On Christmas Eve children tie their stockings to the wagon wheels and then in the morning discover six inches of new snow has fallen.

They fight off depression with church services, singing, and dancing. The stone affords them a good dance floor, the fiddles come out, and music floats out over the emptiness of the country.

And finally they arrive at the Hole. It is a cleft almost too narrow for a man. The top part of the descent runs at an angle of twenty-five degrees and then comes to a drop of forty-five feet. Below the drop the walls widen but the pitch here is forty-five degrees and it runs for a quarter mile. Then the thing fans out and finally for the last third of the way to the river the wagons faced deep sand. So they blast and hack at the rock with picks, shovels, sledgehammers, and chisels. Men dangle in barrels over the forty-five foot drop as they drill holes for the blasting. The barrels are twisting in the wind, the rope suddenly looks thin, day after day, week after week, there is never enough powder for blasting, all too often muscle must do the work. Food is scarce, no wood for the fire, the barrel and the rope—will it hold?—the drill hot in the hand, the dust, rock chips, bruises, smashed fingernails, sore shoulders. Time is running out. In fact, time has stopped, the barrel is twisting in a time warp. But supplies are shrinking, they must press onward. Swing the hammer, hold the chisel, dangle and stuff in the powder. This is a mission. The storms roll in, the temperature drops to zero. A second drop of fifty feet, which is lower down in the gap, is avoided by chiseling out a narrow ledge. Men work on this ledge on a rock face of fifty degrees and when they've finished they have created a scant suggestion of a road with a kind of guardrail, a solution named Uncle Ben's Dugway after its inventor, Benjamin Perkins. And of course, down below by the river's banks men are building a ferry for the wagons.

For six solid weeks they hack and blast.

On the evening of January 25, 1880, they are finished.

IT IS SEPTEMBER OF 1870 and John Doyle Lee is meeting with his father by adoption, the Prophet Brigham Young. Lee has been finally enjoying the fruits of labors—a fine big brick house, herds of fat cattle, fields of bountiful crops. He is nearly sixty, and living as a fabled patriarch surrounded by his numerous wives and yet more numerous children. There have been increasing protests from Mormons and non-Mormons that the church is sheltering the men who murdered at Mountain Meadows. Pressure is building to bring to trial all those involved.

Brigham Young says, "I should like to see you enjoy peace for your remaining years. Gather your wives and children around you, select some fertile valley, and settle out here."

Here is the rock desolation of southern and eastern Utah, a place beyond the reach of laws or prying eyes.

Lee answers, "Well, if it is your wish and counsel."

"It *is* my wish and counsel," the Prophet replies with a firm and sharp tone to his voice.[20]

John Doyle Lee immediately returns home and begins selling off all his property and preparing to move his wives and his herds into an uninhabited wilderness. He alights on the upper Skutumpah valley, about twenty miles south of modern day Bryce Canyon National Park, on November 21, 1870, and what he sees delights his eyes. "This," he notes in his diary, "is on the Most lovely little vallies in the Mountains; abound on either side with rich luxurant feed, fine Springs & Meadoland with several 1000 acres of choicest Farming land, surrounded with beautiful Mountain scenery with alpine tops towering above the Ever green Cedars & shuberies. On the East & west are low roling ranges, forming the appearance of a Plain. At [this] point I intend locating a Portion of My Family. All who have passed this way fell in love with location."[21]

And then, without warning, he is notified that he has been excommunicated from the church he has faithfully served these many years. It is made retroactive more than a month, to October 7, 1870. Lee tells his wife Rachel and she reminds him of the dream he had at that time. Ah, yes, he recalls this vision that came to him. A bunch of people rush into his house and they are dressed in rabbit skins and buckskins. Lee is preparing a feast and the strangers seek to take over his home. They accuse him of sleeping with his two daughters. Lee declares his innocence. He looks over and sees that two of his wives, the very mothers of the daughters in question, are surrounded by impenetrable brambles. When he reaches out to them, his hands are pricked. He goes outside to get a load of wood and finds an Elder half naked and dirty and struggling to wash off his filth. "The Dream was impressive," Lee realizes, "& in connection with others rested heavily on My Mind. I was aware that Satan was working through certain Persons to inJure Me."[22]

Still, the dream is not enough to take the hurt out of his heart. In early December, he rides off for a private meeting with Brigham Young. He asks his father, "how it was that: I was held in fellowship 13 years for an act then committed & all of a sudden I Must be cut off from this church. If it was wrong now, it certainly was wrong then."

Brigham Young tells him, "I want you to be a Man & not a Baby."

The next day Lee receives an anonymous note: "If you will consult your own safety & that [of] others, you will not press yourself nor an investigation on others at this time least you cause others to become accessory with you & thereby force them to inform upon you or to suffer. Our advice is, Trust no one. Make yourself scarce & keep out of the way."[23]

Then he receives new instructions to go down the Paria and found a ferry where the canyon meets the Colorado. He is to take but two wives. He makes a winter passage, driving his cattle through the quicksand. When he comes upon the mouth of the Paria, one of his wives names it Lonely Dell. He begins to build a home in the place Father Escalante named a century earlier as Get Out If You Can. He is now in the heart of the stone.

NO ONE IS SURE which wagon went down first. Such matters are always subject to debate for Americans. But the party moves the bulk of the wagons down and across the first day and manages the rest on the next day. Joseph Stanford Smith and his family are the last. He has been working inside the notch all day helping with ropes to keep wagons from breaking away and flying down the route to certain destruction. He is later working to help with the ferry when word comes that all the wagons are down. But he knows this is not true because his family and his gear are still up on top. He drops his shovel and clambers back up the notch.

He finds his wife Arabella huddled on some old quilts on hard packed dirty snow. Her baby is in her arms.

She says, "Stanford, I thought you'd never come."

"But where are the other children and the wagon?"

"They're over there," she says. "They moved the wagon back while they took down the others."

Stanford gets very angry, throws his hat on the ground and stomps on it.

"With me down there helping get their wagons on the raft, I thought some one would bring my wagon down."

"I've got the horses harnessed and things all packed," she reassures him. They go to the wagon and find the horses hooked to the doubletree; one of them, old Nig, is tied to the rear axle. Their fourth horse died at Fifty Mile Camp. The children crawl out of the wagon and their father hugs them. He unlocks the brakes, his wife beds the baby down inside the wagon and Stanford drives the team toward the Hole In The Rock.

They get out and look at the first section, a drop of one hundred fifty feet. He says, "I am afraid we can't make it."

"But we've got to make it," she says.

"If we only had a few men to hold the wagon back we might make it, Belle."

"I'll do the holding back on old Nig's lines," she says. "Isn't that what he's tied back there for?"

"Any man with sense in his head wouldn't let a woman do that."

"What else is there to do?" she asks.

"But, Belle, the children?"

"They will have to stay up here. We'll come back for them."

"And if we don't come back?"

"We'll come back. We've got to!"

She puts the three-year-old on a quilt a little distance back from the notch and puts the baby between his legs.

"Hold little brother 'til papa comes for you."

She tells her older daughter to sit in front of her brothers and say a prayer.

Now, they are ready to go into the hole.

Stanford sits in the wagon and braces his legs against the dashboard, Belle wraps Nig's lines around her hands. The first lurch down the steep incline almost pulls her off her feet and she races to keep up with Nig. Then the horse rolls to one side and gives a neigh of terror. She thinks, "His dead weight will be as good as a live one." Her foot gets caught between two rocks, she kicks and frees it, but loses her balance and topples over after old Nig. Sand blinds her, rock rips her flesh. The wagon slams against a huge boulder and this blow jerks her back up on her feet and then throws her against a cliff.

The wagon stops. They have survived the hard passage from the top. Her face is white, and she is dirty and bloody.

He leaps down from the wagon and asks, "How did

you make it, Belle?"

"Oh I crow-hopped right along!"

Nig is lying on the ground more dead than alive. Stanford looks up the chute and sees some cloth stuck on a rock. He looks at her and sees a pool of blood forming on the rocks at her feet.

"Belle, you're hurt! And we're alone here."

"Old Nig dragged me all the way down."

"Is your leg broken?"

She kicks him in the shin with all her might.

"Does that feel like it's broken?"

Old Nig gets up, his legs trembling, his hide bleeding. Stanford climbs back up the notch and fetches the children. And they all continue on their journey. They're only halfway there.

AN END IS APPROACHING for John Doyle Lee and he knows it. He has had nineteen wives (once marrying a mother and her three daughters) and fathered sixty-five children. He is the adopted son of Brigham Young, a man he still reveres, and now he is an outcast from his own church. Because of his excommunication his wives began to melt away, seeking life with men who are not tainted. Lee bitterly resents this. When he first visited one of his homes and felt the scorn of his wives that dwelt there, he refused to enter and slept out in a wagon in the rain until their hearts turned and they beseeched him to enter. But that is behind him now. He lives in his Lonely Dell with his memories of that long-ago day in Mountain Meadows. He hears talk of a plan to establish a mission on the upper San Juan (what will become the Hole In The Rock expedition) and thinks he will join this movement. Though he is old, he is still fierce. He has created a new ranch single-handedly, crossed into Arizona, and explored the Hopi and Navajo lands. He is restless.

And then on September 24, 1874, a federal grand jury sitting in Beaver City, Utah, names him along with some other men for the murder of at least fifty men, women, and children on September 16, 1857 at a place called Mountain Meadows. Lee is arrested sometime later on a visit to one of his families. He is found hiding under some hay. He is not worried. He does not think a Utah jury will ever convict him. He is tried in July, 1875, and the trial results in a hung jury. Lee is sent to prison to await a second trial, which is held in September, 1876.

He sits through testimony that he knows is untrue—accusations, for example, that he cut the throat of a teenage girl that day. He does nothing, perhaps, because he knows what is true and he cannot say the truth. And perhaps because what he saw there and did there that day is a burden he can no longer carry.

The wagon train of westward immigrants is from Missouri, and they pass through Utah at a bad time. Since Utah became a territory, largely settled by Mormons, the settlers have ignored the federal government and its officials and followed leaders of the church. Now, an American army is crossing the plains to put down what is seen as a rebellion. And other settlers are moving west. For months, church leaders have been whipping up feelings among the faithful in Utah to fight to the death for Zion. This is the atmosphere the wagon train from Missouri enters. And the members of that party do not help the situation. They are heard boasting of the killing of Joseph Smith, and they denounce the Mormons.

The slaughter happens in southern Utah, at a place called Mountain Meadows. First, Indian allies of the Mormons attack the wagons and then John Doyle Lee arrives to try and calm the tribesmen. He also persuades the party to surrender to the Mormons and is said to have guaranteed their safety. The Missourians put down their arms. Mormon troops are present and one Saint lines up by each Missourian.

The men are shot dead; the women and children, except for the very youngest, are clubbed to death or shot. Sixteen or seventeen children under the age of seven survive and are adopted into Mormon families. The number of people in the wagon train is never determined. Fifty? A hundred? They are tossed into mass graves, and then beasts dig them up, and that is why bones lie bleaching all over the meadows. The property of the immigrants is taken to Cedar City and auctioned by the church. Lee helps with this task. Everything is done to obscure the event.

No one today can even be certain of the exact date: it has been repeatedly shifted, probably to distance the slaughter from known communications with church leaders. The role of Mormon commanders that day is also obscured.[24] Finally, only one man is truly left, John Doyle Lee, who sees his role in the drama very simply: "and I the scap[e] goat." He notes that he wept when the massacre occurred, and it is true that the Native Americans present that day afterwards always called him "Yawgatts," Cry-baby.

He is condemned to death and has a choice of being hung, shot, or decapitated. He selects death by the gun. They take him back to Mountain Meadows; he rides in a wagon seated on his coffin. The governor has let it be known that if Lee will but confess and name the others, his sentence will be reduced or possibly commuted. He says nothing. As they form up the firing squad on the killing ground, Lee walks about the meadow that no longer exists. Huge herds of cattle have grazed it flat, and floods in 1861 and 1873 have eroded it into a gravelly waste. Minutes before his execution Lee hands over a letter for his wife Rachel:

> Morning clear, still and pleasant. . . . Since my confinement here, I have reflected much over my sentence, and as the time of my execution is drawing near, I feel composed, and as calm as the summer morning. I hope to meet my fate with manly courage. I declare my innocence. I have done nothing designedly wrong in that unfortunate and

lamentable affair with which I have been implicated. I used my utmost endeavors to save them from their sad fate. I freely would have given worlds, were they at my command, to have averted that evil. I wept and mourned over them before and after, but words will not help them now it is done. My blood cannot help them, neither can it make the atonement required. Death to me has no terror. It is but a struggle, and all is over. I much regret to part with my loved ones here, especially under that odium of disgrace that will follow my name; that I cannot help. . . .

To the Mothers of My Children

I beg of you to teach them better things than to ever allow themselves to be let down so low as to be steeped in vice, corruption and villainy, that would allow them to sacrifice the meanest wretch on earth, much less a neighbor and a friend, as their father has been. Be kind and true to each other. Do not contend about my property. . . . Live faithful and humble before God. . . . Remember the last words of your most true and devoted friend on earth, and let them sink deep into your tender aching hearts. . . .[25]

His sons watch from horseback. Lee is given a chance to speak and he says, "I have but little to say this morning. Of course I feel that I am on the brink of eternity. . . . I am ready to die. . . ." They have Lee sit on his coffin. He is blindfolded, but his hands are left free. He asks the riflemen not to mutilate him and to please aim at his heart. And then they pump five rounds through his body. He falls into his coffin without a sound, only the twitching of his left hand indicating that he is dying. His sons haul the body away. The family buries John Doyle Lee in his temple robes. In April 1961, he is reinstated to full membership in the Church of Jesus Christ of Latter-Day Saints.

THE WEEKS ROLL BY and then the months and they gather up a string of place names that mark their camps. There are Register Rocks, Cottonwood Hill, Little Hole In The Rock, Cheese Camp, Gray Mesa, Clay Hill Pass, Cow Tank, Dripping Spring, Harmony Flat, Grand Flat, Snow Flat, The Twist. "It's the roughest country you or anybody else ever seen," notes Elizabeth Morris Decker. "It's nothing in the world but rocks and holes, hills and hollows. The mountains are just one solid rock as smooth as an apple."[26] The weather goes hard with them. One day it is two feet of snow: the next axle-sucking mud. Wagons are abandoned, stock gives out and sometimes heifers are hooked up to pull the loads. As they near Elk Ridge a cedar forest forces them to chop a way through its thickets.

Then they came up against Comb Ridge, a sandstone monolith with walls a thousand feet high, and they plod along against its stone flank toward the San Juan river. They hope that at the river they can round Comb Ridge and ride along the San Juan. But they find the river has knifed through, leaving sheer cliffs. There is no way around the endless stone ridge. They begin to build a dugway up the smooth skin of the rock, and the work takes days. And then they go up.

Each wagon has seven spans of horses for the ascent. Some of the horses go down on their knees struggling for a foothold. On the worst sections the men have to beat the animals. Some of the horses go into spasms and near-convulsions. The route is marked by blood and matted hair. They spend the first three days of April in this hell for man and beast. Once on top, they drive their wagons along the ridge for a few miles and then drop down into a wash where only few small dugways have to be built.

They are on level ground now, only twenty miles from their goal, the new settlement at Montezuma. But there is almost nothing left of them, not of the men, or the women, or the children, or the oxen, the horses, the cattle, the mules. They camp near the river just east of Cottonwood Wash. Only eighteen easy miles now to their goal, some of it on an actual dirt road. They cannot do it. They have been six months making two hundred and sixty miles. There have been babies born, but not a single life lost. Not one. They have suffered, to be sure, but they have also danced and sung and laughed and heard that fiddle music hit the rock walls. And they've prayed, prayed real prayers. They are on a mission, they know this and it has given them the strength to pull through. They've gone through some of the roughest country on God's good earth, done it with wagons to boot, and they have maintained simple decency. No shootings, no knifings, no cannibalism in the hungry times. Truly, they have kept the faith.

They settle at the base of Comb Ridge, call their community Bluff, and begin laying out house lots and fields. They dig irrigation ditches, throw small dams across the river. Floods come and tear the hell out of everything. Rebuild. They are on flat ground and they tell each other flat ground tastes sweet to them now. They've been through the country and it has taken their measure. But they have avoided the Escalante, at least they can say that.

Escalante Canyons

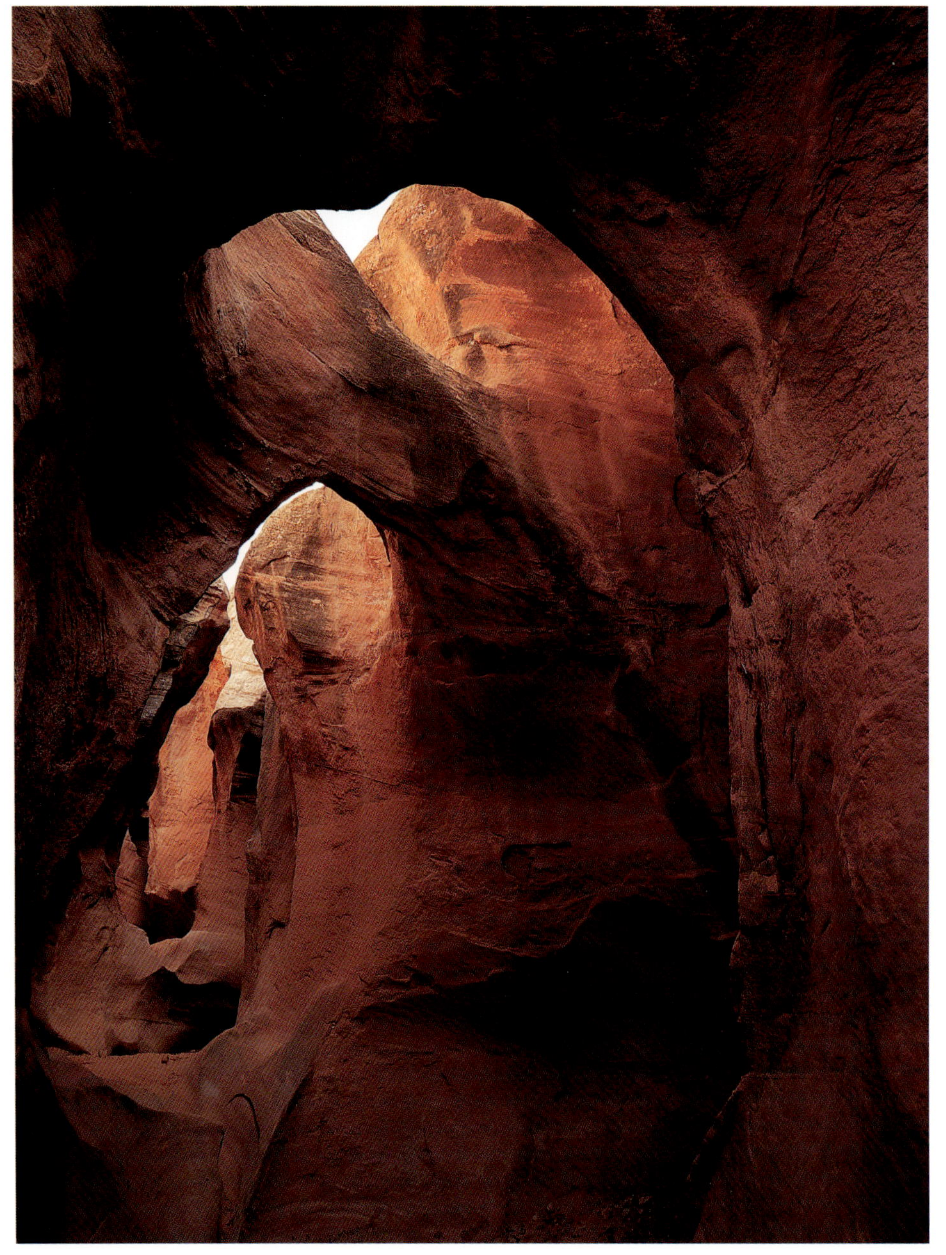

At Peek-a-Boo Canyon, natural bridges arch over the Navajo sandstone slot canyon. (Escalante Resource Area, Wilderness Study Area, BLM, Utah)

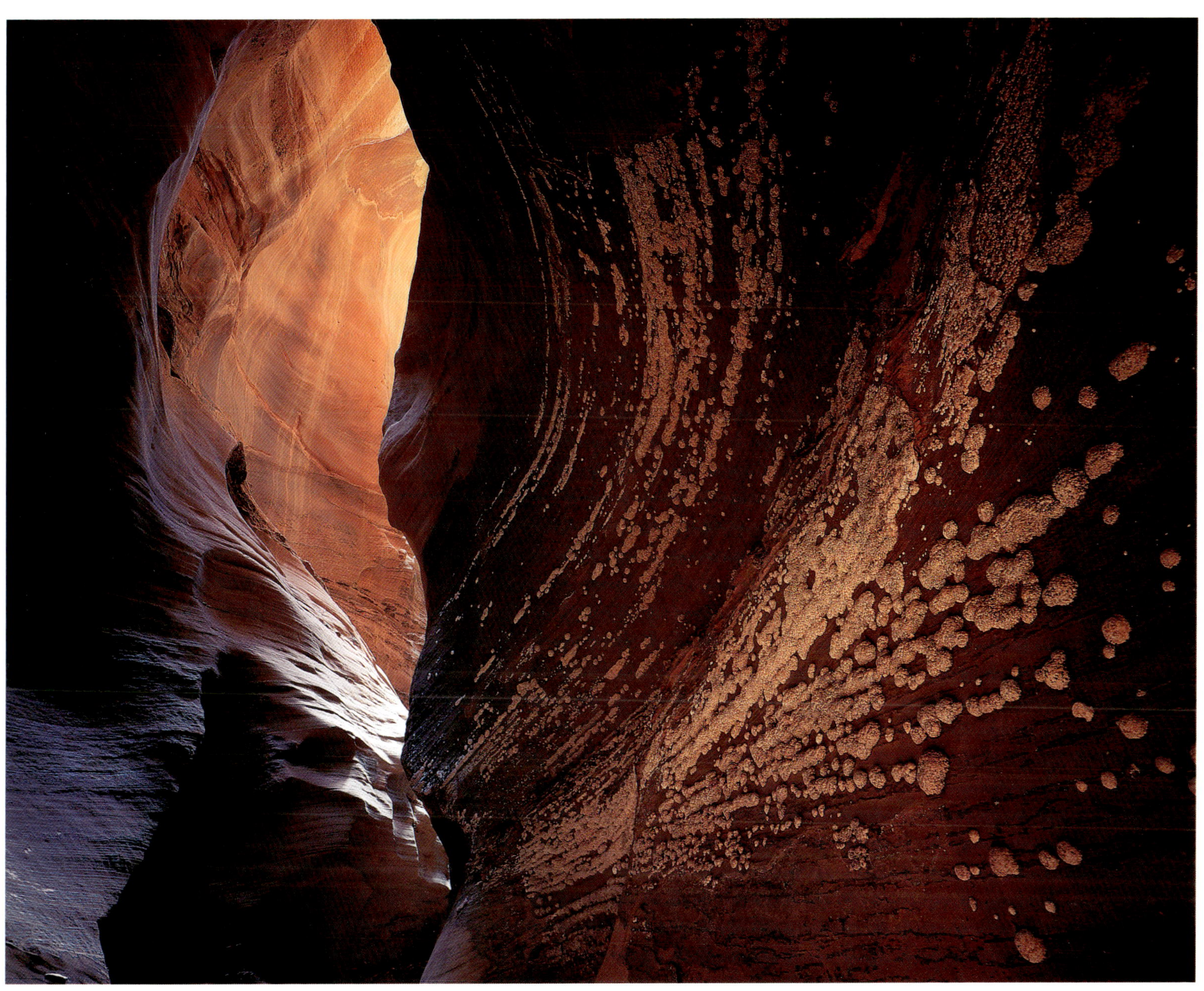

Calcium carbonate dots the Navajo sandstone wall of this slot canyon.

(Escalante Resource Area, Wilderness Study Area, BLM, Utah)

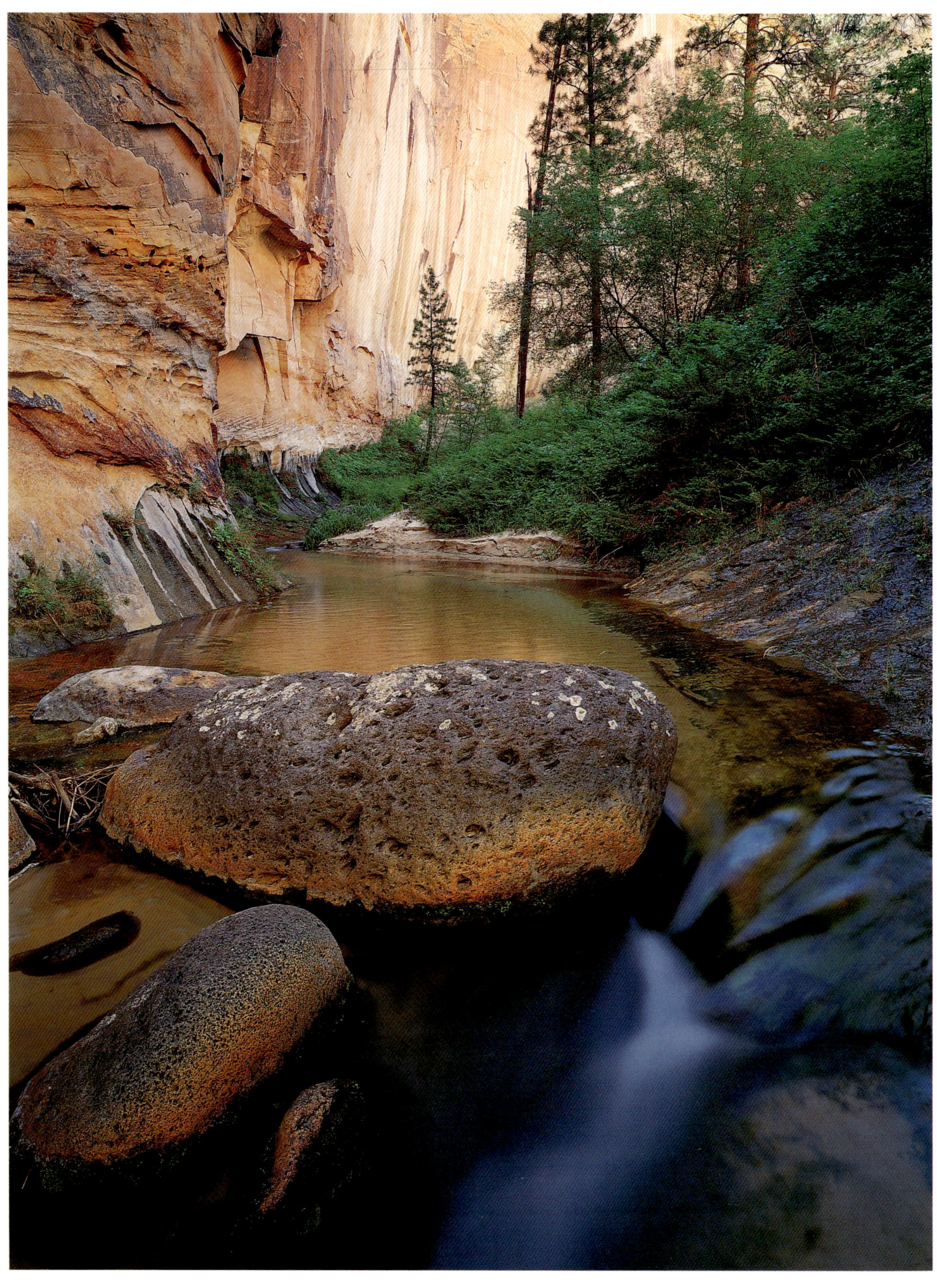

Canyon walls clad in maidenhair ferns (Adiantum capillus-veneris) *and lined with ponderosa pines* (Pinus ponderosa) *overlook boulders in the stream. (Death Hollow, Escalante Resource Area, Wilderness Study Area, BLM, Utah)*

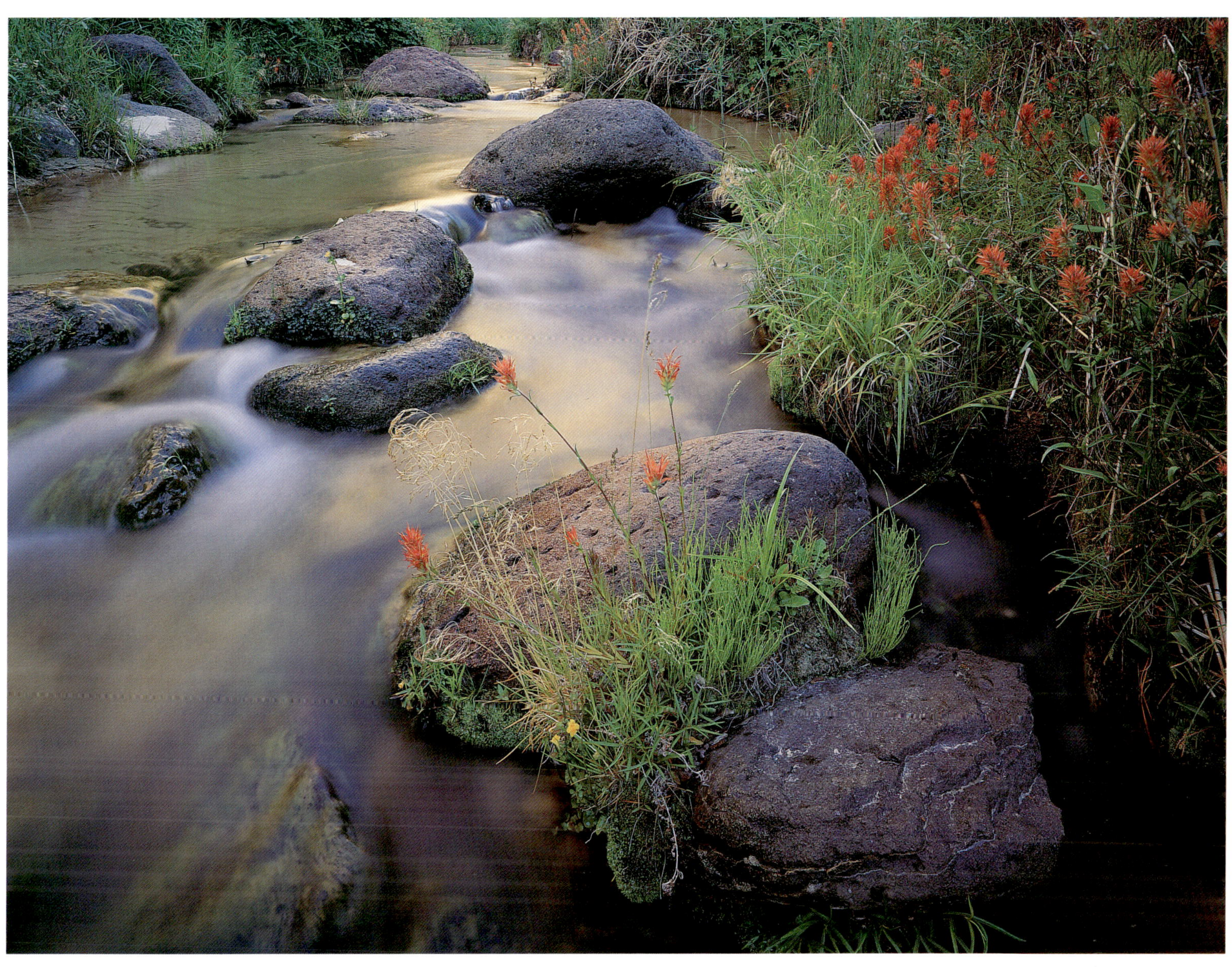

Flowering Indian paintbrush (castilleja sp.) *and yellow monkeyflowers* (Mimulus guttatus) *flourish in a boulder-strewn stream. (Death Hollow, Escalante Resource Area, Wilderness Study Area, BLM, Utah)*

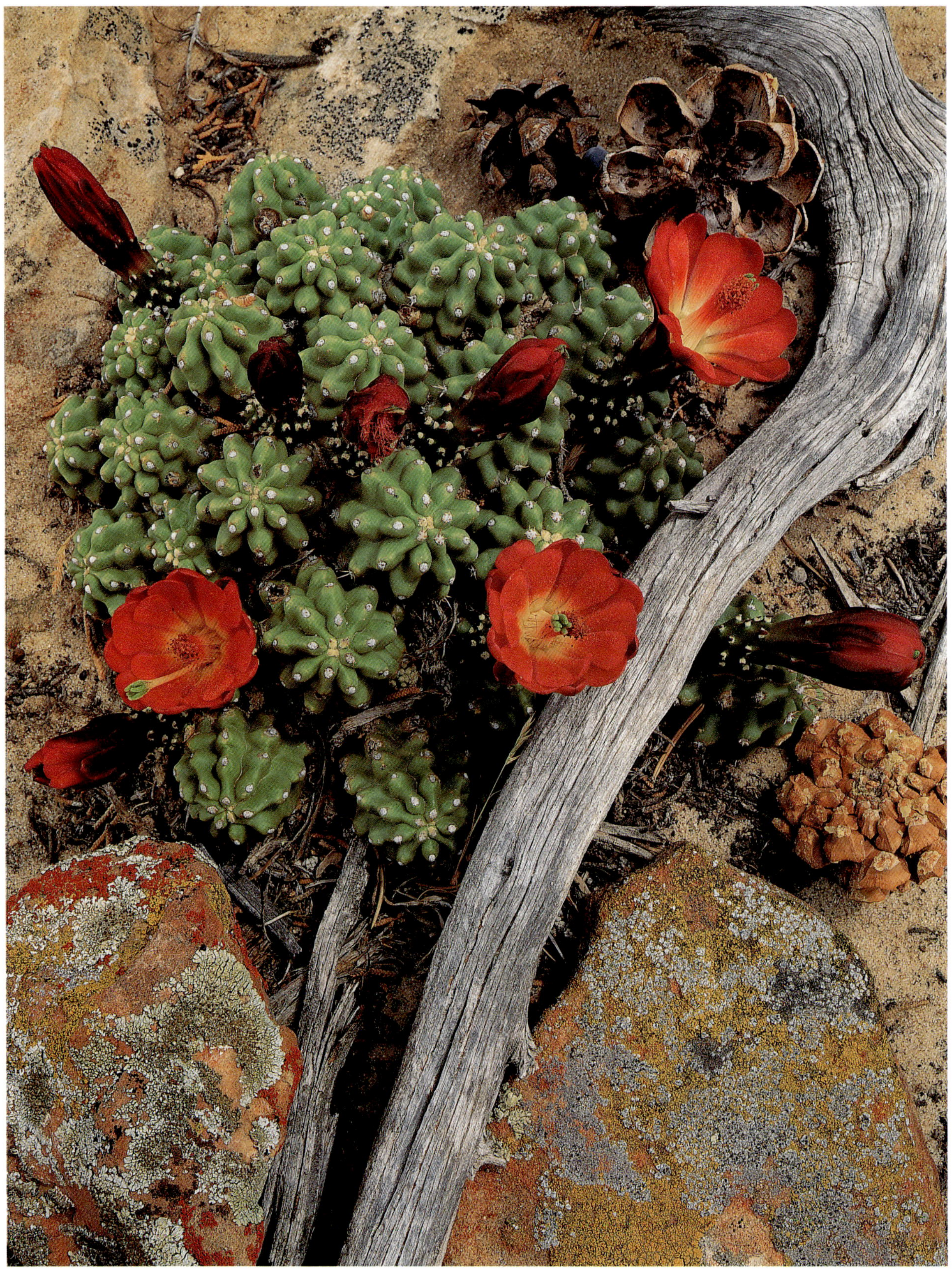

Claret cup cactus (Echinocereus triglochidiatus), *pinon pine boughs* (Pinus edulis), *and lichen-covered rocks. (Near Death Hollow, Escalante Resource Area, Wilderness Study Area, BLM, Utah)*

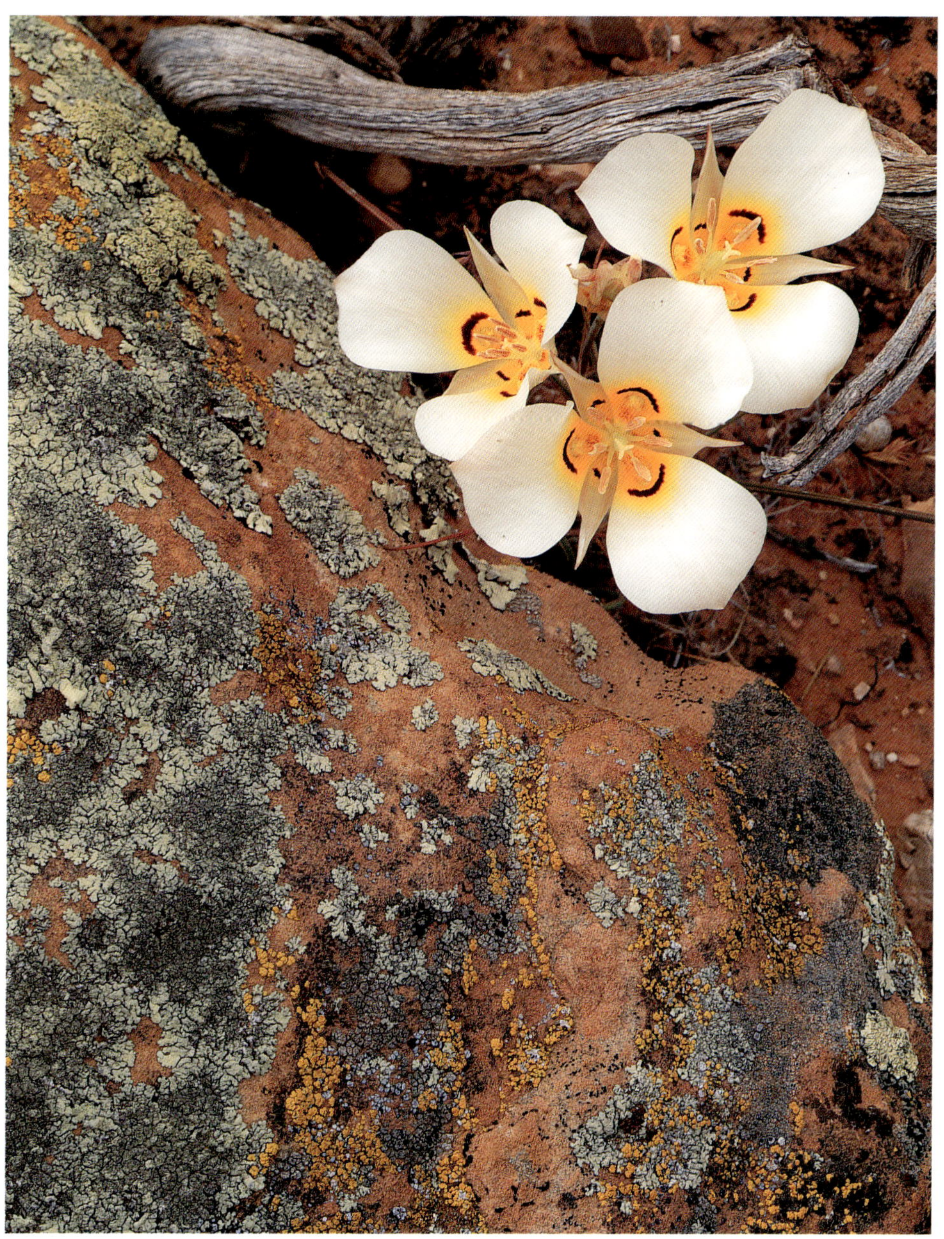

Sego lilies (Calochortus nuttalii) *flower amid lichen covered sandstone. (Davis Gulch, Glen Canyon National Recreation Area, Escalante Canyon Complex, Utah)*

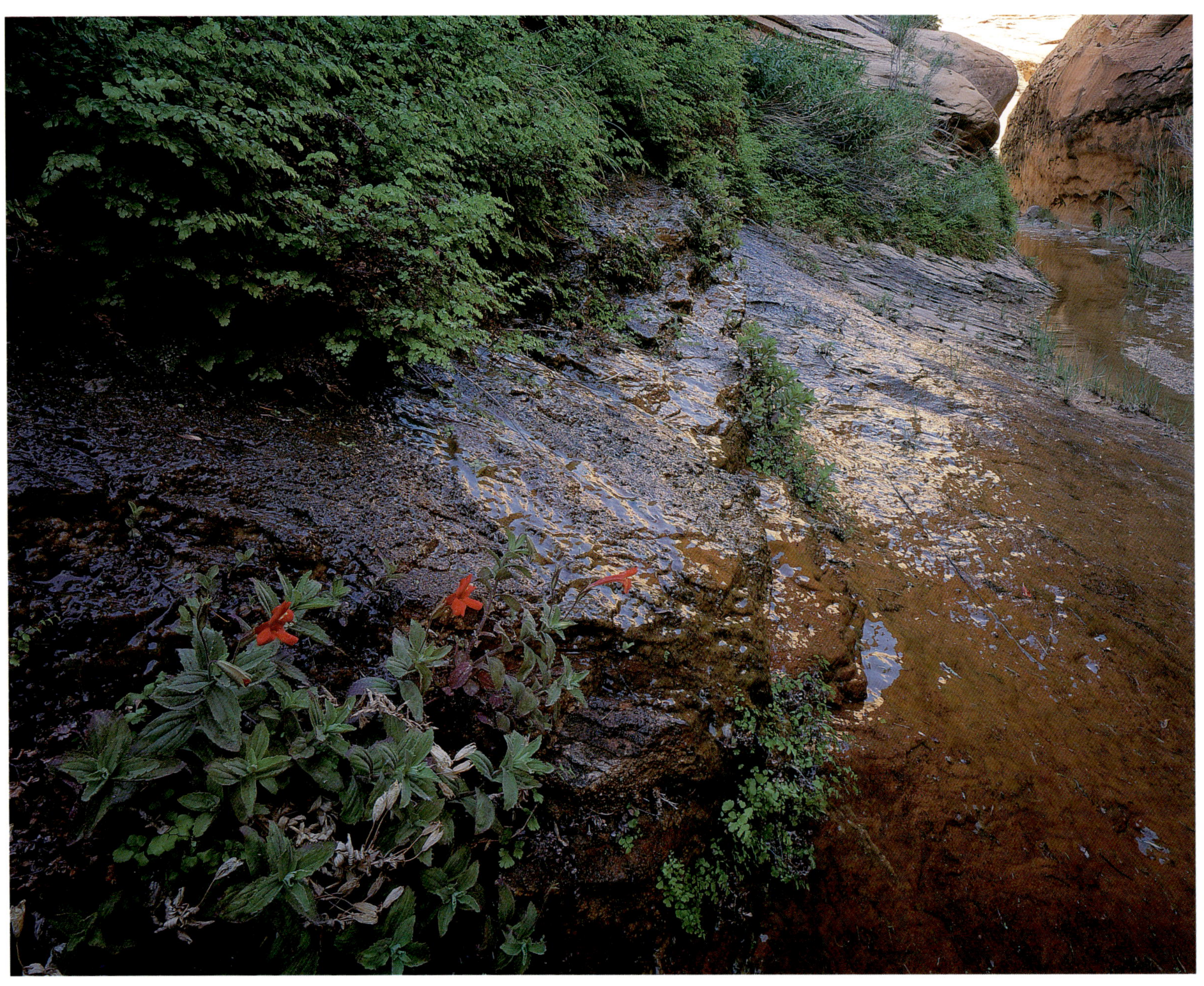

Scarlet monkey flowers (Mimulus cardinalis) *and maidenhair ferns find nourishment at a seep. (Escalante Canyon, Glen Canyon National Recreation Area, Utah)*

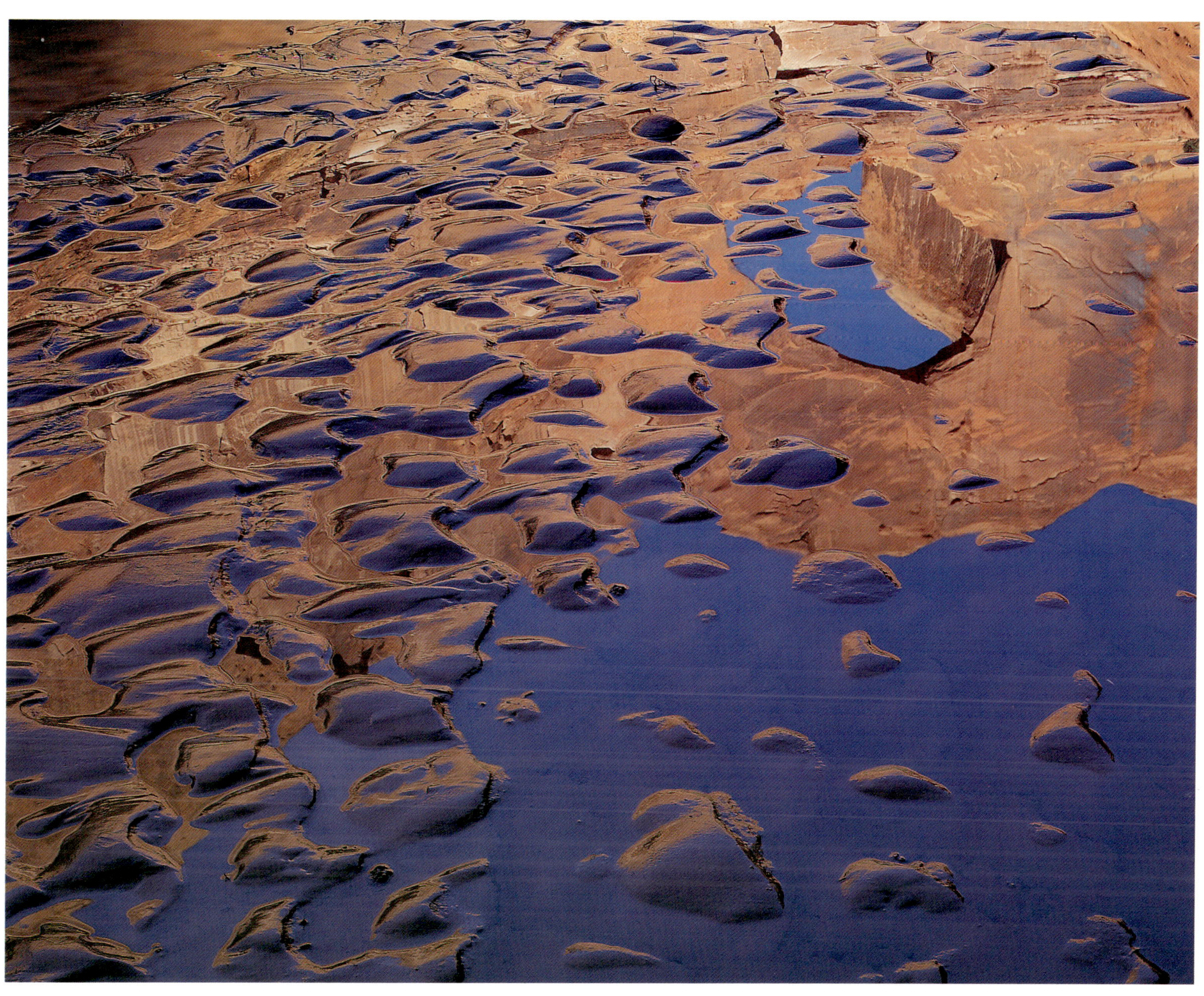

Stevens Arch is reflected in sand bar ripples. (Escalante Canyon, Glen Canyon National Recreation Area, Utah)

Sand sunflowers (Helianthus anomolus) *lie half buried in the sand in an Escalante side canyon. (Escalante Resource Area, Wilderness Study Area, BLM, Utah)*

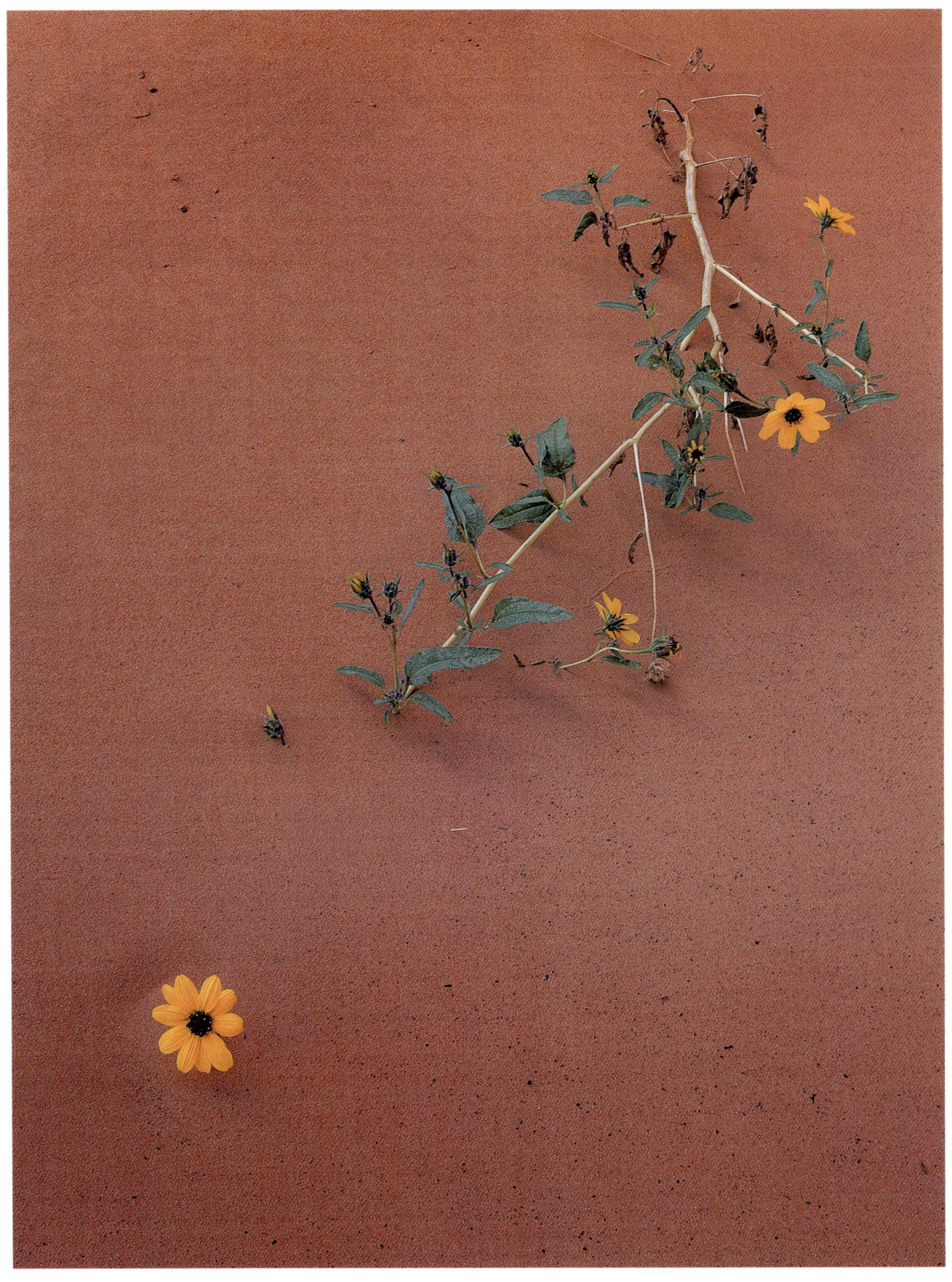

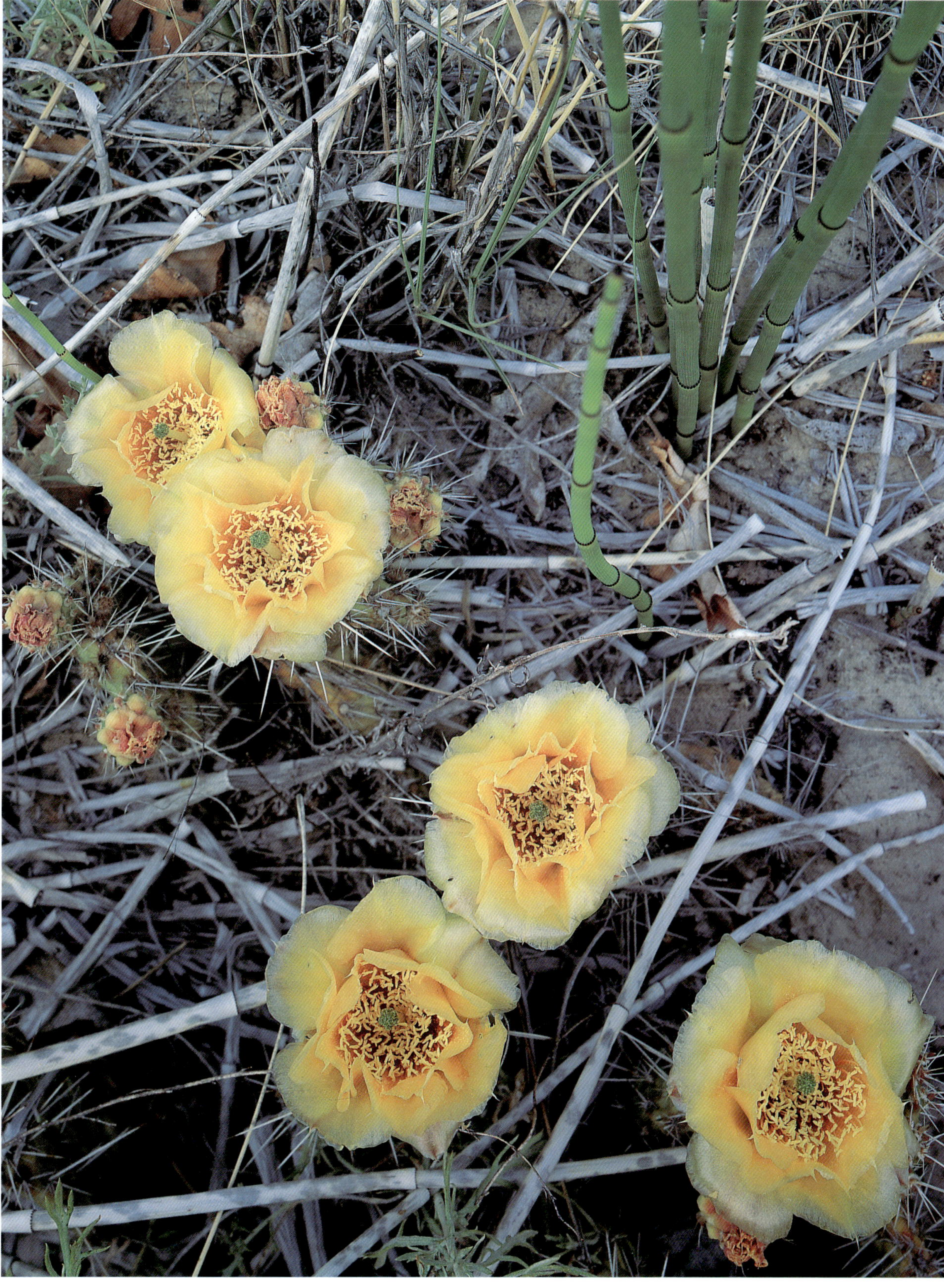

Prickly-pear cactus (Opuntia sp.) *blossoms among horsetails. (Death Hollow, Wilderness Study Area, BLM, Escalante Canyon Complex, Utah)*

Flowering lupine (Lupinus sp.) *and scarlet bugler* (Penstemon eatoni) *against canyon walls. (Little Death Hollow, Escalante Resource Area, Wilderness Study Area, BLM, Utah)*

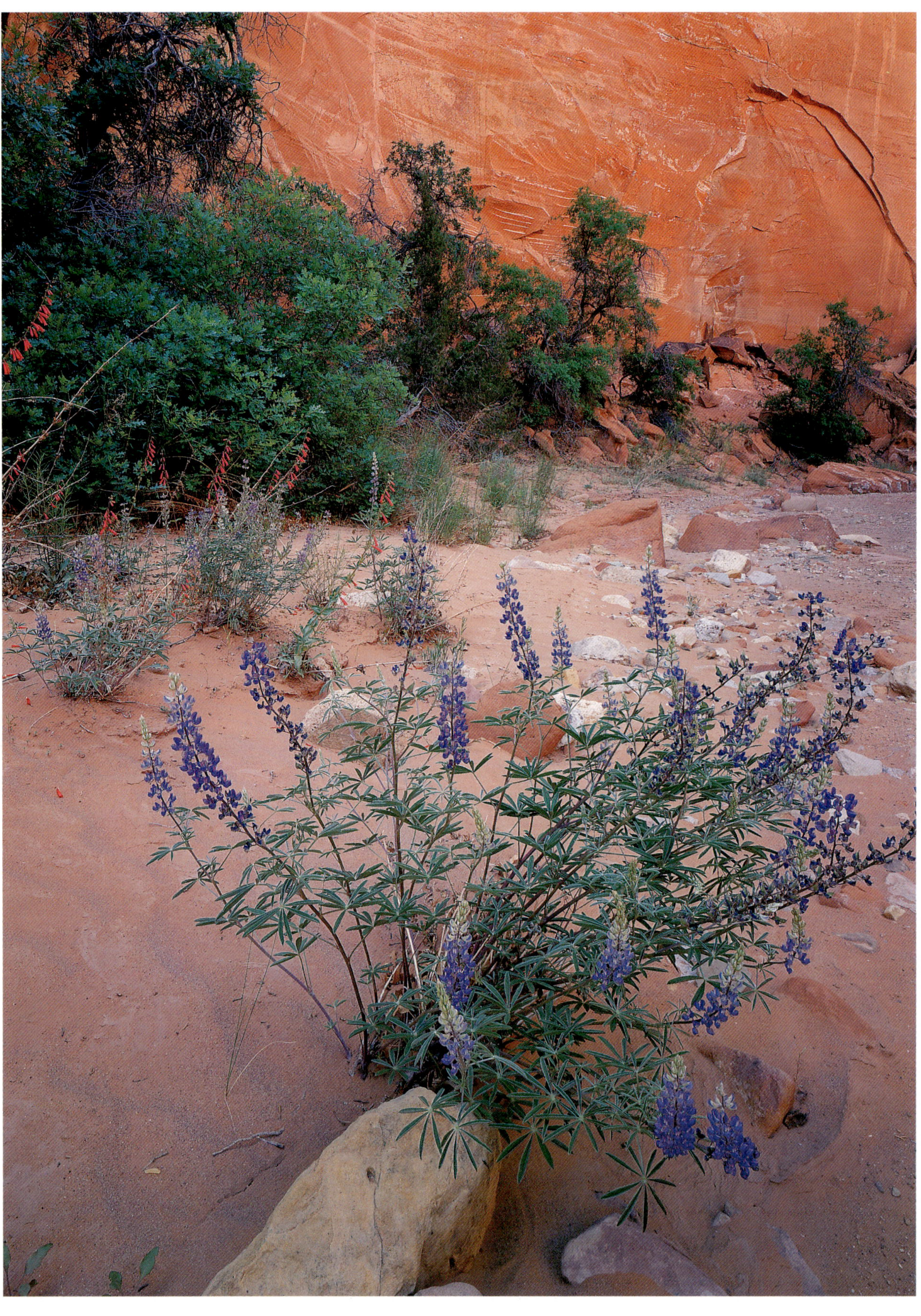

THE IDEA HAD floated like a leaf over the region since the 1920s. First had come Zion National Park and Bryce Canyon National Park, and then, by presidential proclamation, six national monuments. Decades after the efforts to settle southern Utah stalled, this carving up of the ground into parks seemed like the best business of all. The Great Depression brought Utah in general, and southern Utah in particular, to its knees. Measures such as the Civilian Conservation Corps were lifesavers. There was talk of a new five-hundred-seventy square mile park at the junction of the Colorado and

New Dreams

Green Rivers, an idea that Utah's congressional delegation thought wonderful. Harold Ickes, the Secretary of the Interior, was on a roll by the mid-thirties. He envisioned a Grand Teton national park in Wyoming, Kings Canyon in California, Olympic in Washington state. And in Utah, the department conjured up Escalante National Monument, a plan that would embrace two hundred miles of the Colorado drainage, running basically from the current site of Glen Canyon dam upstream well past the junction of the Green and the Colorado. It would lock up 6,968 square miles, eight percent of Utah.[27] This massive set aside would displace 463 families, 144,000 sheep, 26,000 cattle and 2,600 horses. Actually, the people would not so much be displaced as cut off from free economic use of the land. The creation of the park would not condemn towns, but rather give them a new source of income other than grazing on terribly overgrazed land.

Clearly, the plan would declare the past a failure, a past where each family required fifteen square miles to barely eke out a living. That was part of the pain and the problem. The park—constructed almost totally out of federal land—meant that the American way of wresting a livelihood from the ground had broken down in this empire of stone. It would also mean giving up any fantasies about the future, any dreams of a diamond as big as the Ritz, of the Big Rock Candy Mountain, of oil wells hiding under the rim rock, of the gold mines that someone could hope to stumble upon some day, of the big dams that would magically transform life and make everyone rich and all the days easy. The plan was kicked around for five years, whittled away by politicians (by 1938 it was a trim 2,450 square miles), and killed in 1940 when the government turned its face toward war, and shelved ideas of parks, and leaped at hopes for massive dams to power the emerging military-industrial complex.

Today, it haunts us as the path not taken, the trifling investment of our treasure that would hardly have disturbed the petty cash drawer of government and would have created a global wonderland. A tiny fragment of this vast vision is lodged in Canyonlands National Park, a petty tract of five hundred fifteen square miles set aside on September 12, 1964. Of course, some of the things in the Escalante plan of the thirties are at the moment beyond our reach—things like the one hundred ten murdered canyons and ten thousand Anasazi sites that now sleep in the mud under the stagnant waters of Lake Powell, the tub created by Glen Canyon dam. Today, the same ground hosts about 28,000 people and they continue to live largely in the traditional economic world of rural poverty—a place where high school graduates take the first bus out of town. In a sense, the Escalante plan is still possible. The only real barrier remains the same: ourselves.

We have to wonder about these dreams we have, dreams which seem to possess and rob us of our senses. We like to blather about private property, state's rights, stewardship, and a growing economy. It is difficult to comprehend what these terms and ideas have to do with canyonlands and deserts of southern Utah. Private property has hardly ever penetrated the area—no one is willing to pay taxes on most of it. The land is almost totally federal, not state, so maybe we should talk about federal rights. Stewardship is hardly a fit topic for conversation in a region that for a century has lived off federal dole (subsidized grazing, subsidized mining, federal water projects, and so forth). Until quite recently, we have engaged in and enjoyed brutal overgrazing and devastating mining practices. Economics are another humorous issue—our way of doing business has clearly failed in this region. We have been on an almost endless Hole In The Rock expedition and continue to dream of towns that will not thrive in places best left alone. *Salsipuedes,* Get Out If You Can. The priest sensed something we should pay attention to: this place is a trap for our normal schemes and our typical appetites.

But, like creatures in a fable, we have a second chance at life. We can still set the ground aside and skip our foolish talk about rights and money and progress and growth. The very rock itself mocks this language. We know looking into this enormity of stone and time that we are a passing thing, hardly a tick on the giant geologic clock that confronts us. We have the possibility of creating a kind of temple ground, a place without a clear religion or creed but with the space and silence to feed all those spiritual impulses that roll within us. Within the language of the 1964 Wilderness bill millions of acres can be secured from our failed ambitions. Within the plans of the Southern Utah Wilderness Coalition, we can find a ready blueprint—one perhaps too modest in its goal but certainly a giant step forward and on the right track.[28]

I want to forget for a moment the way we talk, this world of board feet of lumber, hiking permits, grazing permits, mining claims, recreational values, off-road vehicle routes, river permits, hunting licenses, catch limits, and regulated primitive campsites. Forget the Lycra, gear ratios, aluminum frames, Cordura nylon, polypropyhelene, waffle stompers, lens, film stocks—all the f-stops terminate here. Go, go to the color; pink dripping down the wall, white flowers of columbine booming up from the deep green mat of leaves, rock strata tilting, water gushing through slots and holes and falling off ledges and idle in quiet pools reflecting cliffs, of mud dry and cracking, stone columns, bare rock ridges, melting hillsides bereft of plants, willows licking the water's edge, a dead mouse frozen in a final leap across the smooth trackless sand, the wind coming up biting the face with cold, eating at the eyes with dust. Cloud roll in and the sun flees, everything going dark, and then a downpour chilling the night and becoming snow. Walk, climb, throw away the map, ignore the place names, just move, but slowly, down the canyon, through the slot, under the arch, looking, but looking for nothing, just being, and being is everywhere. No list, no checklist, no life list, no death list, no words, no names. Silence. The leaves are rustling, a bird calls, water spills. Forget, forget lists, names, maps, goals, itineraries, photo ops, view points, guides, forget it all. No peak bagging, no canyon bagging, no route bagging. No trophy hiking. No purpose, please God, no purpose. And no special places, never special places—that way of thinking is going to be the death of us all. This is it. We have made it. We are nowhere.

We will resurrect the vision of our forebears in the 1930s—the initial plan for an Escalante National Park. Only, having learned from experience, we will offer some refinements. This will be a national park with no roads—pavement will end in the parking lots of the park's borders. There will be no concessionaires, no hotels, no little signs telling you what you are seeing. Rangers will keep a low profile, perhaps being dressed in a tasteful camouflage, with their faces blackened out for night work, and there will be no campfire sings led by park employees. Ideally, no one will ever go there—a park without visitors. But we must be reasonable in a country of a quarter billion people and in a world that is all but bursting at the seams. We will allow Homo sapiens to wander about the vast tract. Of course, there will be no typical search and rescue units. All such work will be handled by carrion eaters, vultures, skunks, coyotes, armies of ants, billions of microorganisms, and such rescue as might avail will be piecemeal. There will simply be this big, empty thing, which small, quiet signs on the borders of the expanse will designate as a nuclear test site or some other fashionably forbidden zone. Mountain bikes will be outlawed and their introduction into the park will be a capital offense. Of course, the same will apply to off-road vehicles, helicopters, hot-air balloons, motor boats, and any other forms of transport except for the human leg or non-motorized river craft. This will entail absolutely no sacrifice on our part since the nation is already studded with parks where machines roam freely and feed greedily on the air and the earth. Ah, yes, one more detail: cellular phone usage will be promptly punished by garroting—second offenses will be treated more seriously.

No map will ever indicate where this park is located. People will just bump up against it like all those mysterious castles hidden in the forest that populate our fairy tales. They will then abandon their cars, toss away their keys and identity papers and disappear into the land and into themselves. No one will be allowed to publish a word about the place—brothers and sisters, this is the final communiqué. Photographers, naturally, will be shot out of hand. Eventually, it will exist as a rumor, an Atlantis lost beneath the waves of our busy republic. Like the black budget of our intelligence services, no financial clue in government documents will reveal the existence of the place. All this can be done. Good heavens, if we can fight secret wars, as we have, we can have a secret national park. We are dealing here with known technology. For years we have had military bases in this country which neither appear on maps nor whose existence is ever admitted by our government. Clearly, any reasonable person must admit that these Pentagon installations are not nearly as vital to our national security as a secret national park.

I realize that my modest and sketchy proposal will be attacked by radical elements as inadequate, cowardly, and perhaps even despicable. But I am not a purist, and prefer to be seen as a practical man susceptible to compromise. While wild-eyed revolutionaries might bellow and demand (a word I hate) say fifty percent of Utah, I am, in my reasonable way, seeking only eight to ten percent of the state. And, at that, merely a portion of the commonwealth that has produced little or no wealth on balance and failed to nurture or shelter many of the state's citizens. Clearly, no one in their right mind or not incapacitated by strong drink or illicit drugs can fail to see the moderation in leaving ninety percent of the land to the vices, follies, and appetites of one singular species (Homo sapiens), and setting aside a trifling eight or ten percent for the rest of God's creation. This is a standard of giving set by the most respectable religions: tithing. So let us get on with this modest and just act.

We will create that thing we have always professed to abhor. A void. A wasteland. A wilderness screaming in the night as a full moon slides across the sky. It will really have no value in dollar and cents. But it will help pay off our true national debt. For a while it will bear the brunt of industrial tourism and this will bring in more of our beloved money than the livestock and mines can. But this boom, like all booms, will have its limits and its end. The global tourism of today is a byproduct not of ecological insight, but of ecological wantonness—the gobbling up of our fossil fuel reserves and the looting of the forests and minerals and sea beds of the undeveloped world. It will eventually come to a shuddering end—and that end will occur in a century or much less—and then the jet-powered-backpacking world will vanish in the twinkling of an eye. This is not necessarily a terrible thing. I doubt God really made the world so camcorders would have a place to function. Or so that rapacious citizens of industrial societies would have room for locating their second homes.

In the end, the world will return to a form of travel that long ago characterized it: pilgrimages. The canyons will bring on the pilgrims, foot weary folk seeking things they cannot name in this heart of stone.

We will do it. The only question is when.

The answer is within us right this moment and we just have to finally say it out loud. We have run out of schemes. It is time to let our dreams float free and take us into the country.

They have already laid down the law for us, a statute whose language comes in good part from the pen of Wallace Stegner. We should be law abiding, I believe, and repeat over and over this simple thought:

> In order to assure that an increasing population, accompanied by expanding settlement and growing mechanization, does not occupy and modify all areas within the United States and its possessions, leaving no lands designated for preservation and protection in their natural condition, it is hereby declared to be the policy of Congress to secure for the American people of present and future generations the benefits of an enduring resource of wilderness.
>
> The Wilderness Act of 1964

We are all together now. The Native Americans, the padres, John Doyle Lee, the Saints, the Hole In The Rock expedition, all of us. We are in this place, get out if you can, but why try? We gather around a small fire in the evening, the stars are out, the wind has died, the coals fall and there is an occasional crackle. We can all say it now, say it together.

Amen.

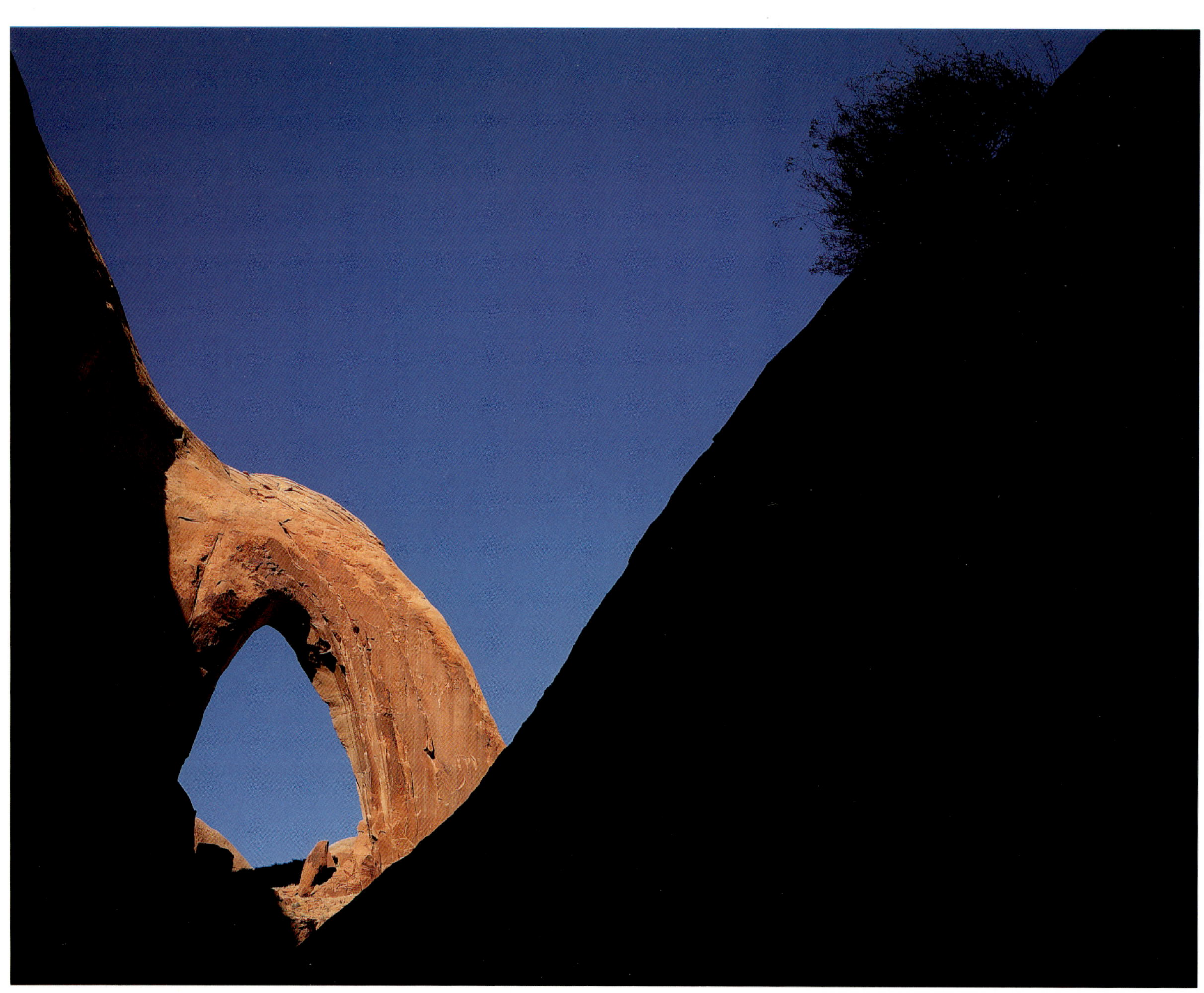

Broken Bow Arch rises above Willow Canyon's Navajo sandstone walls. (Escalante Canyon, Glen Canyon Recreation Area, Utah)

Sand sunflowers (Helianthus anomalus) *bloom against the desert varnished side wall of an unnamed side canyon. (Escalante Resource Area, Wilderness Study Area, BLM, Utah)*

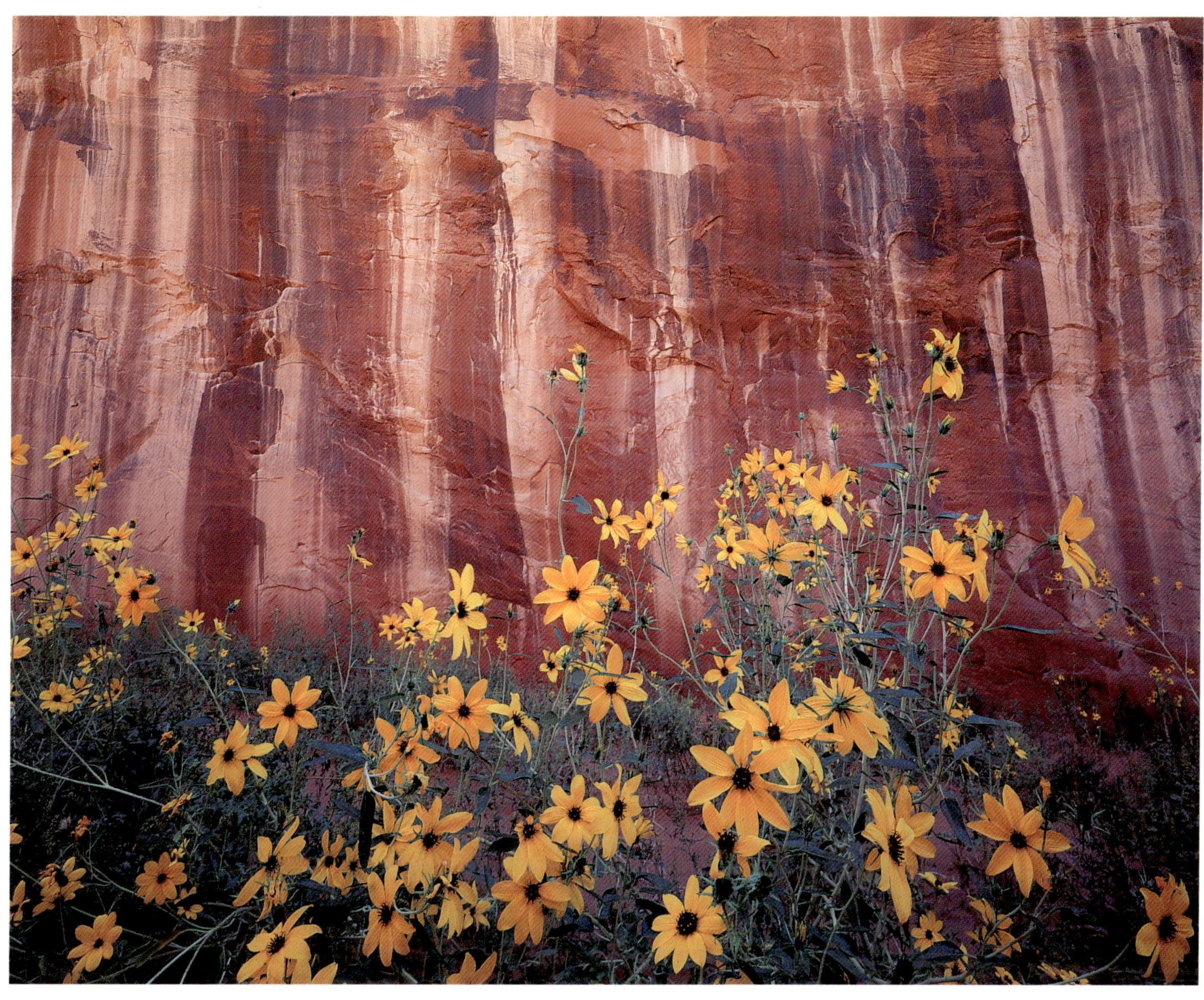

Fremont cottonwoods show their fall color beyond a cascade over sculpted sandstone. (Coyote Gulch, Glen Canyon National Recreation Area, Utah)

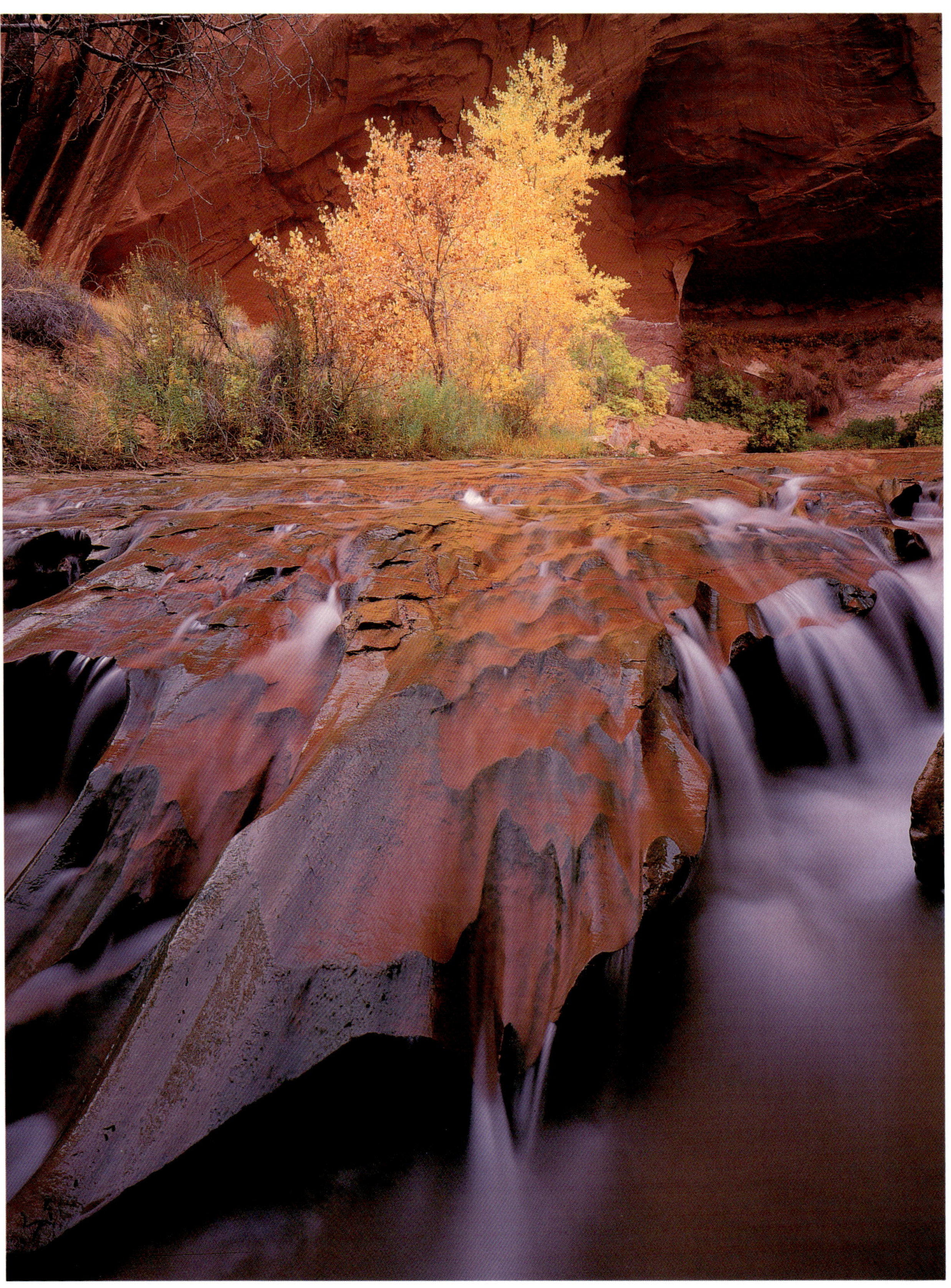

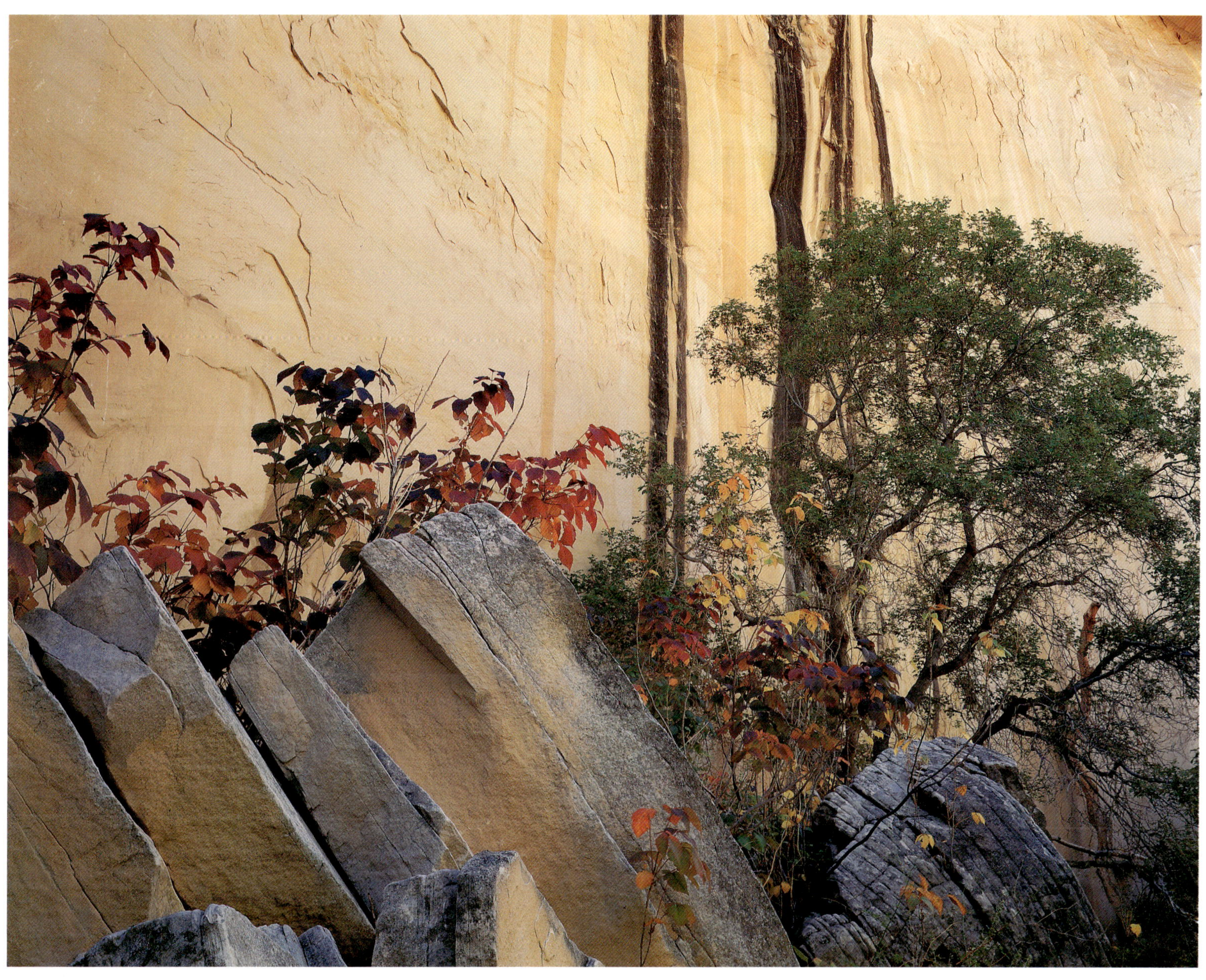

In autumnal hues are Gambel oak (Quercus gambelii) *and poison ivy* (Toxicodendron radicans), *seen against the canyon wall.*

(Death Hollow, Escalante Resource Area, Wilderness Study Area, BLM, Utah)

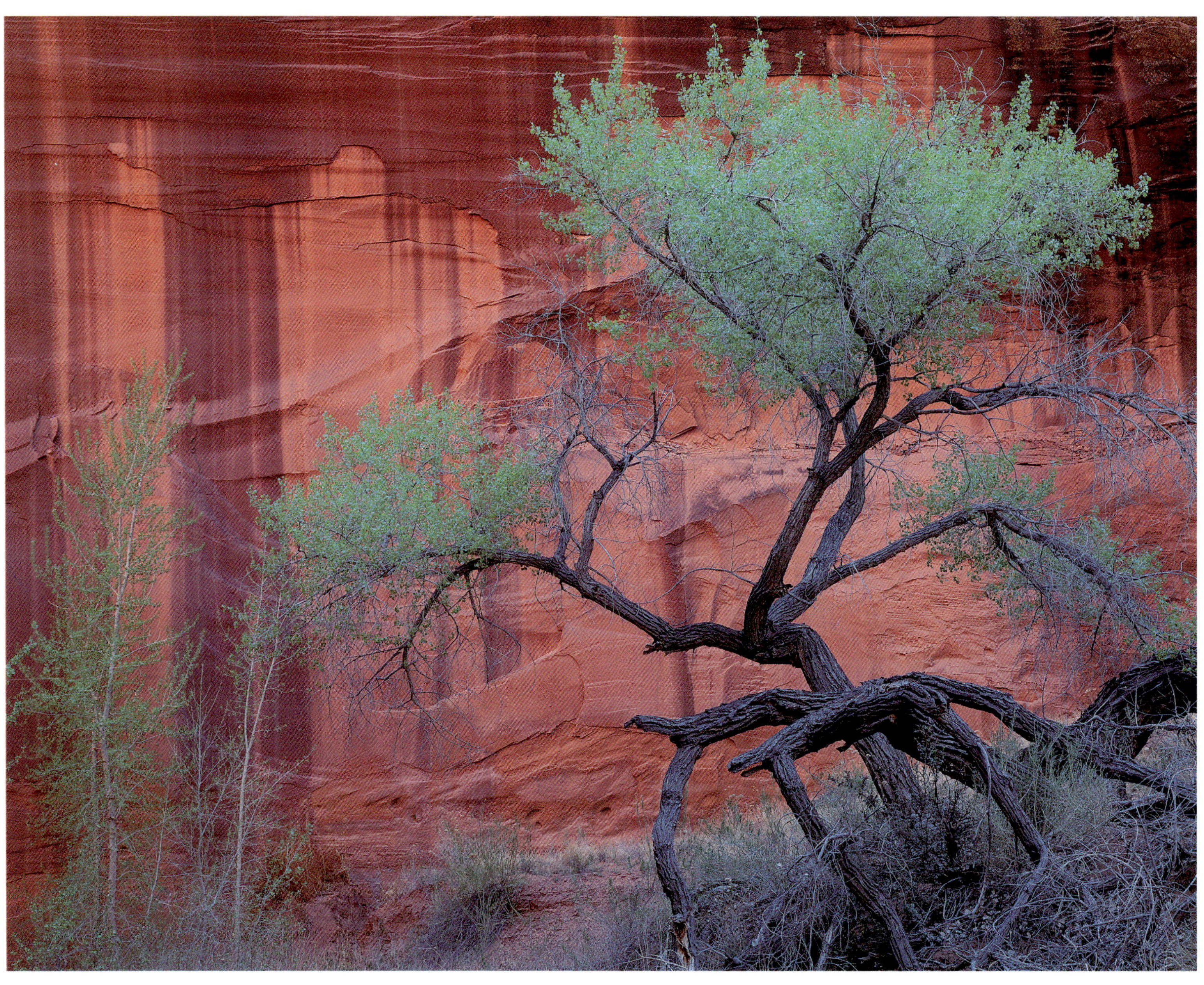

In Neon Canyon cottonwoods line up against the glowing Navajo sandstone walls. (Glen Canyon Recreation Area, Escalante Canyon Complex, Utah)

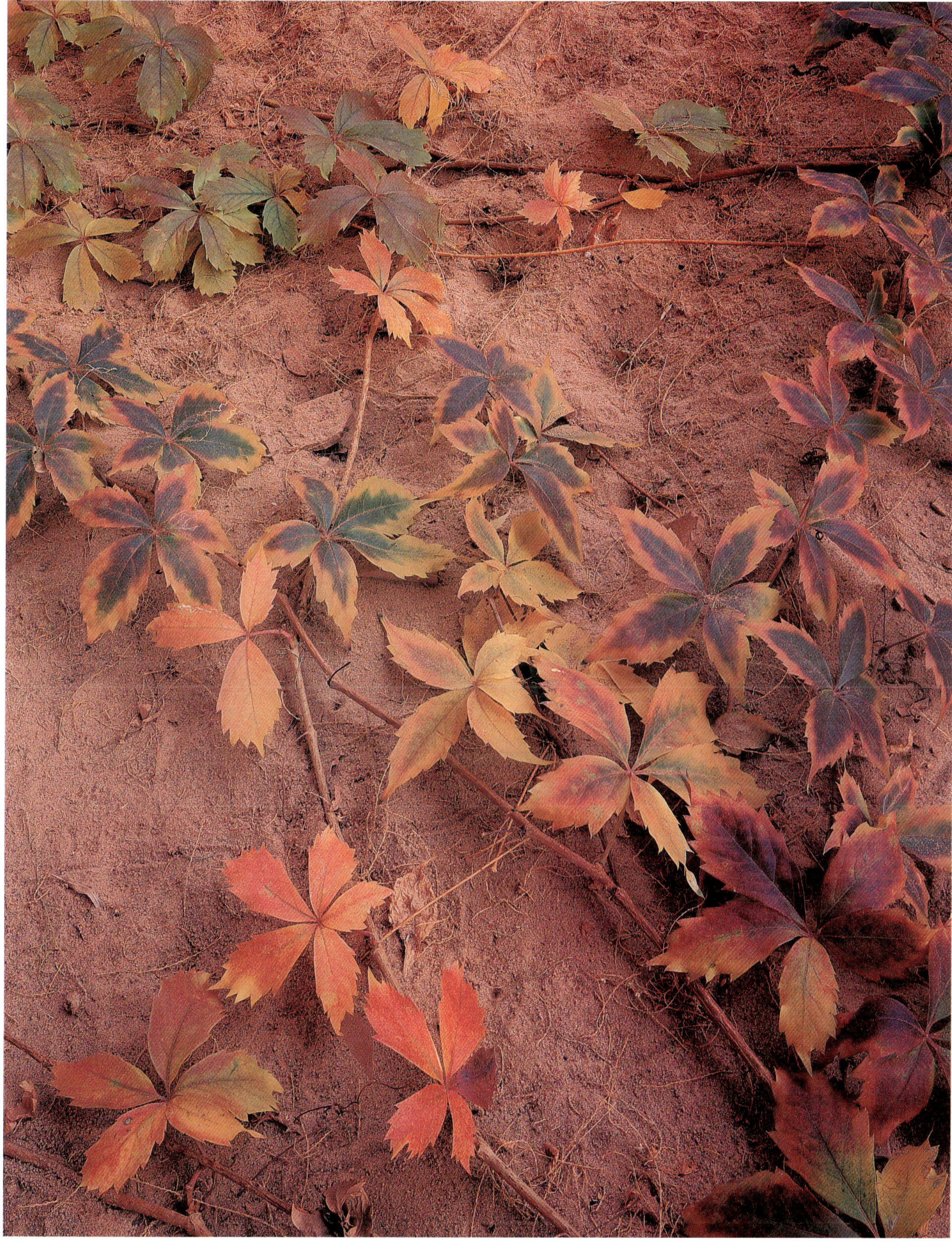

Virginia creeper (Parthenocissus vitacea) *in its fall color brightens the sandy shore. (Stevens Canyon, Glen Canyon National Recreation Area, Utah)*

With its roots exposed, a cottonwood sapling clings to the bank of Willow Creek. (Escalante Canyon, Glen Canyon National Recreation Area, Utah)

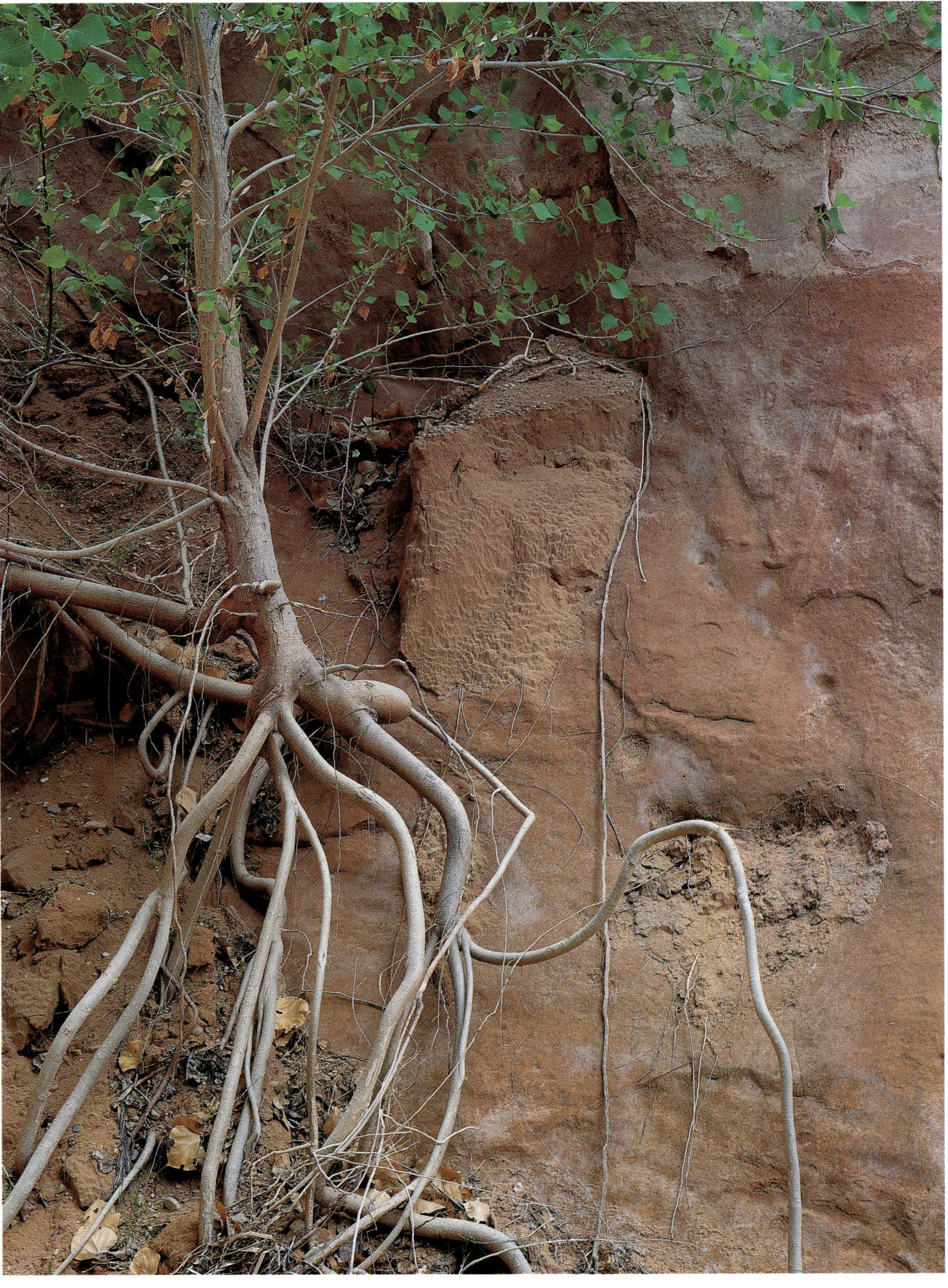

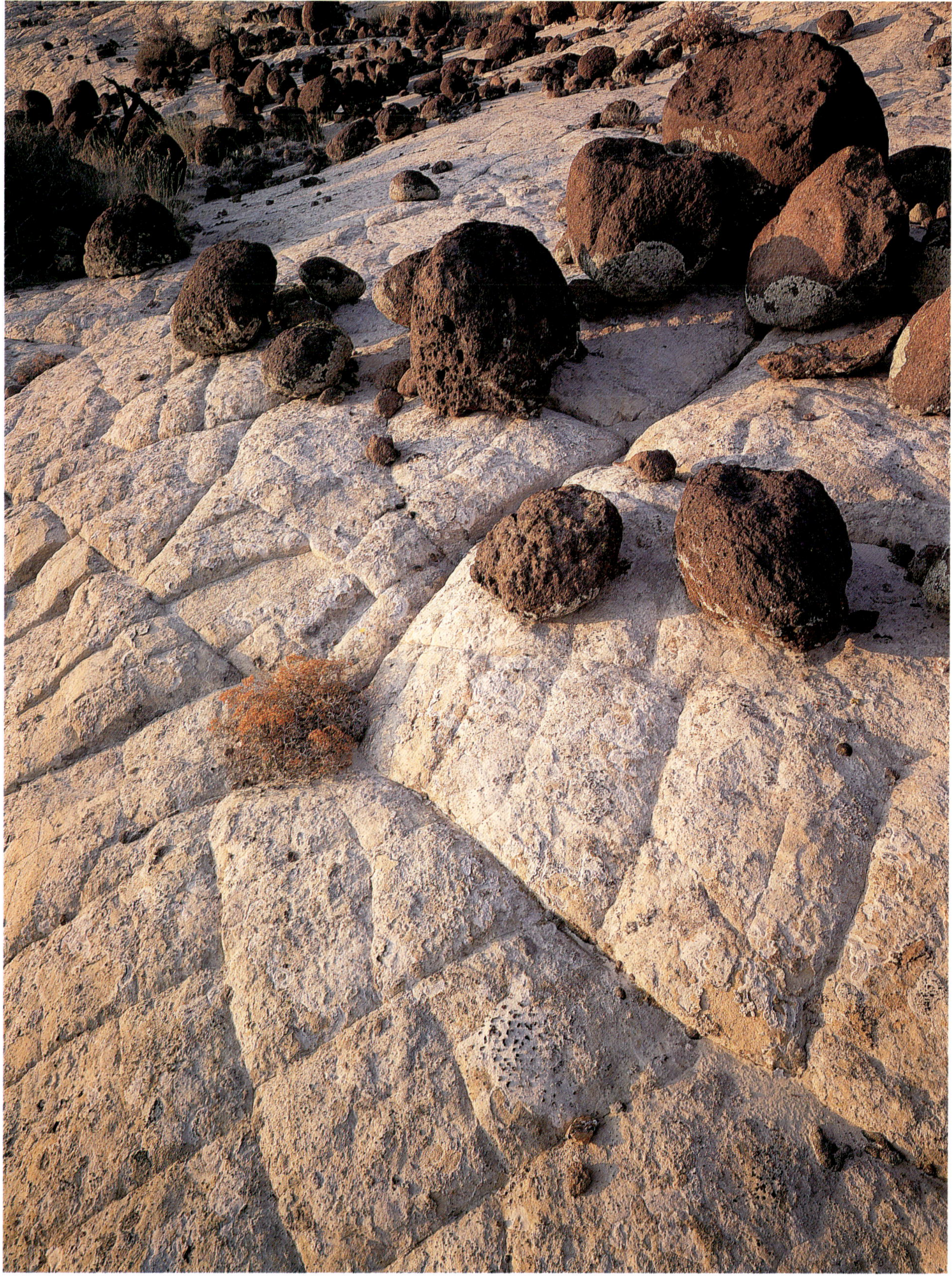

Eroded Navajo sandstone and basalt boulders with wild buckwheat (Eriogonum corymbosum) *growing in the fracture line, near Calf Creek. (Escalante Resource Area, Wilderness Study Area, BLM, Utah)*

Nothing grows here, the grass is gone. Erosion, scrub, bare earth, but not the slightest hint of why anyone ever called it a meadow. Perhaps it is cursed. Probably, bones are still about, but the clumps of hair, they are gone. No faded sunbonnets around to haunt us

Dreams Ending

either. We would like to think it is over. It is all too bloody and painful—the cries of the children, the screams of the women, the blubbering of the wounded and dying men. And, in a real sense, it is brother killing brother, people of one nation finding a way to slaughter each other despite their common language and background and childhoods. At last the volley that cut John Doyle Lee down has died away and the ground is still now.

Lee himself is an obscure figure known to historians and few others. The pain of Mountain Meadows has been so great that few living Americans know that this was the signal slaughter of the pioneer expedition. It

Dreams Without End

comforts me that Lee fled into the world of rock, coursed down the Paria, and found some few moments of peace when the birds would talk to him again and help him send his messages of pain to his loved ones. I think he not only came into the country but came to know the country, and knew it in a way that few nineteenth-century Americans ever permitted themselves. Or few twentieth-century Americans for that matter. We cannot get rid of him. A present without a past means no future. If we are really going to know this country, and do right by this country, and love this country, we must admit all that has happened so that we can honestly dream all that can be.

I think he knew it was beautiful—not minerals, not board feet, not irrigation ditches, not fields, not grazing ground, not mills. Beautiful. There are clues in his endless diaries, small moments about light and rock and wind and sun and . . . birds.

Dreams Beginning

When you go to this ground keep an eye peeled for him, this zealot, this pioneer, this killer, this scapegoat. You might sight him in the canyons. I really think you might. We all need healing, and I believe Paria was the balm for John Doyle Lee, the place where he found the cooling waters to put out the fires raging in his flesh. He'll be polite for a cantankerous man. Don't be foolish and start arguing religion—but, then, any well-bred American knows better than to do such a thing, since we realize religion is a private matter of faith and belief and not a football to kick around in saloons or cafes. But I think you might bump into him; he knew the ground well. And when you do, he'll have clear eyes and I suspect the sun will be on his face.

When you bump into him, ask if he is "Yawgatts?" Cry-baby? I don't think he'll be ashamed of his tears.

THE OLD MORMON HOUSES stand on large house lots along the San Juan river in Bluff, Utah. Originally, each settler got an additional sixteen acres of farmland. Dams were thrown against the San Juan, and repeatedly the San Juan blew them out. Canals silted in, buildings washed away. On December 1, 1883, the church gave the colonists permission to pull up stakes and abandon the mission. The settlers drifted over to Colorado or New Mexico. The church itself did not flourish here: today services are conducted by officials who come down from Blanding.

The route through the Hole In The Rock was abandoned within a year of the migration of the San Juan mission. Now the upper part has collapsed into rubble, and the river crossing itself is drowned under the waters of Lake Powell.

There is a cemetery on a hill above Bluff. The members of the expedition that came through the Hole In The Rock and remained in Bluff now sleep there. The burying ground is stark, gravelly, and there is no grass. Today there are no descendants in Bluff of the people who made up the Hole In The Rock expedition.

Even with all that energy and song and fire and faith, he community could not take root in this country. This is not surprising. There is a history in this ground that can be read by its very absence. The rock has a way of winning and defies our big plans.[29]

For those who love this country, Bluff has a quiet beauty to it. We gave it our best shot here and used some of our strongest and most determined people. And we failed. To sit by down on the banks and listen to the river purr along is good medicine. We did not so much do something wrong as do it in the wrong place. This has become a bad habit with us but I think it is time to change our ways.

Off there, away from the river, is this enormity of stone. People still think they can find dreams out there; a big oil field or gas strike or a second home bonanza. There is a fledgling industry, taking mountain bikers into remote country by helicopter. We are now on the verge of turning the heart of stone into a toy shop. Our special places have become marquees on T-shirts. We advertise property based on the view, the things that we can see beyond the ground we live on, rather than in the ground itself. Our real ground has increasingly become books such as this one. We buy them and hope to possess and own more than we experience. We can read the words, look at the pictures. We can live this lie. The country is not in words, the photographs are cropped clues to the thing itself. Some day, and that day is not so distant, no one will remember the Hole In The Rock expedition or John Doyle Lee or this book. But Paria will be there, narrow and hard and unknowable. And the Escalante will be there, brooding over its secrets and rolling along toward the Colorado. Our decision is simple: don't mess it up, don't think it is a toy, don't think we can possess it and don't miss the chance to have this ground teach us. Here, to experience the place is to be brought to our knees by the force of the country. It is rough as a cob, larger than our imaginations and pursuing a path we are not likely to understand any time soon.

We can do whatever we wish for a while. There have been plenty of escapades in this land of canyon and quicksand and rock and sun. But consider the results. Nothing really sticks here. The railroads do not get built, the mines play out, the ranches burn out the ground and then wither themselves, the communities start bold and then recede into sorrow.

Lincoln was right: "A nation may be said to consist of its territory, its people and its laws. The territory is the only part which is of certain durability." We should listen to birds. Old John Doyle Lee did. And look to the territory ahead, the only part of us that can last or matter.

> This is an automobile age, but I do not have much patience with people whose idea of enjoying nature is dashing along a hard road at fifty or sixty miles an hour. I am not willing that our beautiful areas ought to be opened up to people either too old to walk, as I am, or too lazy to walk, as a great many young people are who ought to be ashamed of themselves. I do not happen to favor the scarring of a mountainside just so that we can say we have a skyline drive. It sounds poetical, but it may be an atrocity. . . .
>
> I think we ought to keep as much wilderness area in this country as we can. . . . We ought to resolve all doubts in favor of letting nature take its course. In a field where nature is preeminently the master artist, where nature can do much more than we can do with all our cleverness, with all of our arts and with all of our best efforts, we cannot improve but can only impair if we undertake to alter.
>
> Secretary of the Interior
> Harold L. Ickes,
> February 25, 1935[30]

Acknowledgments

Anyone who seeks to learn more about the Colorado Plateau learns to be grateful to the libraries of this nation. The government surveys of the latter part of the nineteenth century are but the beginning of a treasure trove of material on this rocky ground. It is not simply that there may be more books about the canyon lands than there are people inhabiting the largely empty region—though this fact is remarkable enough—but the passion that infuses these works, whether they be geological, anthropological, ecological, nature writing, pioneer memoirs or what-have-you, is a unique testament to the power of this ground. For once, we are not dealing with the world we lost—the virgin forests of the East, the prairies of the Midwest, the short grass and tall grass Great Plains, the vast savannahs of California's central valley, well, just about anywhere else in this nation—but with a world that still persists, though we often are afraid to acknowledge this fact, since it might morally impel us to back off and let well enough alone. So a tip of the hat to all those men and women who have gone before us and left stirring records.

In plumbing this rich horde of books I want to particularly thank Ken Sanders, a native son of Utah, who was a fine guide into the rock, and who was generous with his knowledge of both the land and the traces of the land that have graced so many fine volumes. I especially recommend the trailblazing tome he edited, *Utah, Gateway to Nevada.*

Charles Bowden

I wish to thank Terry Bendt for the best focusing cloth in the world and for handing me holders of film as I hung from my cliff-edge perch.

Thanks to Don and Joyce Bayles, Steve Allen, Ginger Harmon, Barry and Celeste Bernards of Escalante Outfitters for showing the way. To Bob Treherne for helping shoulder the loads. Thanks to Ted Nakagami of Fuji Film for express shipping the new Quickload holder to me and to Photographic Works for their high-quality film processing.

A special thanks to poet-photographer Ulrich Schaffer for his unselfish sharing of ideas, vistas, and soggy camps in pouring rain.

Janice Emily Bowers and Robert H. Webb of the U.S. Geological Survey's Desert Laboratory in Tucson,

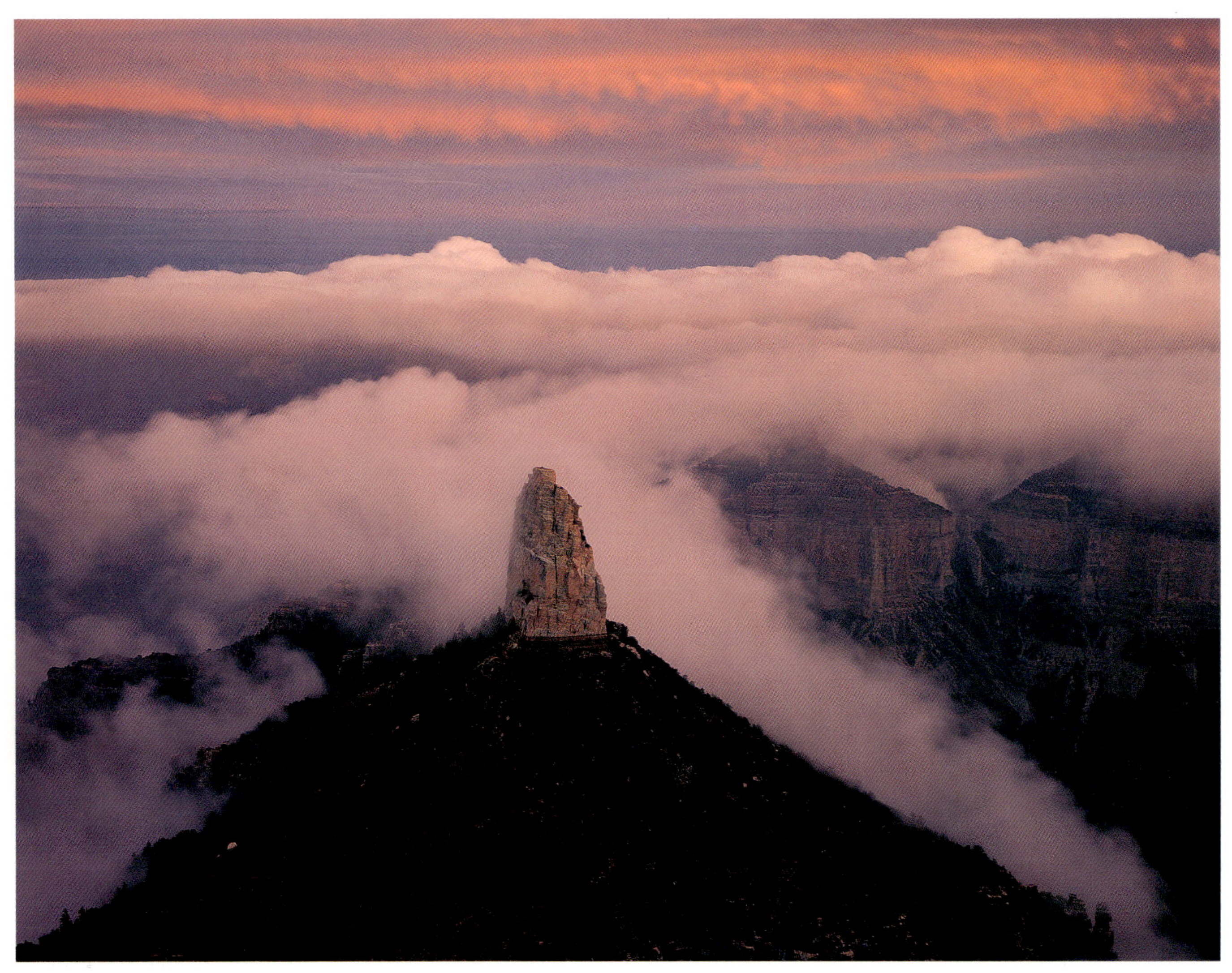

At sunset, Mount Hayden rises through clearing fog at Point Imperial in Grand Canyon National Park, on the North Rim. (Arizona)

Arizona, can even make photographers like me seem credible. Thank you!

The Southern Utah Wilderness Alliance's support of preservation of wild ground deserves not only my thank you, but the collective thank you of an entire nation . . . the real owners of this magnificent canyon country.

Finally, thank you to my son Peter Dykinga and Sandy Smith for keeping my office alive and well during my prolonged time below the canyon's rim.

Jack Dykinga

Notes

OLD DREAMS

[1] and [2] deleted as unnecessary. –Editor

[3] David E. Miller, *Hole In The Rock,* University of Utah Press, Salt Lake City, 1966 [1959], p. 12.

[4] ibid., 8. Morgan Amasa Barton's account of the dangers of southeastern Utah. He was a son of one of the leaders of the Hole In The Rock expedition.

[5] Lee Reay, *Through The Hole In The Rock To San Juan,* Meadow Lane Publications, P.O. Box 640, Provo, Utah, 1980, p. 11. The quote is from the report of Platte D. Lyman on the efforts to find a route to the San Juan country.

[6] LaVan Martineau, *The Southern Paiutes,* KC Publications, Box 15630, Las Vegas, Nevada, 1992, p. 154 passim.

[7] The definitive study of this culture and its triumphs is Wallace Stegner, *Beyond The Hundredth Meridian: John Wesley Powell And The Second Opening Of The West,* University of Nebraska Press, Lincoln, 1982 [1954].

[8] Wallace Stegner, introduction to Clarence E. Dutton, *Tertiary History of the Grand Cañon District,* Peregrine Smith, Inc., Salt Lake City, Utah, 1977 [1882], x.

[9] Juanita Brooks, *John Doyle Lee: Zealot, Pioneer Builder, Scapegoat,* Utah State University Press, Logan, 1992, p. 124. Brooks is that rare and wonderful thing, a balanced, reasonable historian. A Mormon by faith, she wends her way through the controversies surrounding John Doyle Lee and the Mountain Meadows massacre in a manner that inspires confidence. Her work is a model of what benefits history can offer all of us. To read her books is to confront the brutality of our common past and to feel the healing power of truth when it is finally made available to us. Life would be a good deal easier for all of us if more historians explored the past with her cold eye and warm heart.

[10] Robert Glass Cleland and Juanita Brooks, eds., *A Mormon Chronicle: The Diaries of John D. Lee, 1848–1875,* University of Utah Press, Salt Lake City, 1983, V. I., pp.150–51.

[11] Robert Glass Cleland and Juanita Brooks, eds., *A Mormon Chronicle: The Diaries of John D. Lee, 1848–1875,* University of Utah Press, Salt Lake City, 1983, V. II, pp. 99.

[12] Carl Sandburg, *Abraham Lincoln: The Prairie Years And The War Years,* Harcourt Brace Jovanovich, San Diego & New York, 1966, 329.

[13] This account is based on William B. Smart, *Old Utah Trails,* Utah Geographic Series, Inc., Salt Lake City, 1988, pp. 11–25.

[14] C. Gregory Crampton, *Land Of Living Rock: The Grand Canyon and the High Plateaus: Arizona, Utah, Nevada,* Peregrine Smith Books, P.O. Box 667, Layton, Utah, 1985, pp. 185–89.

[15] Robert Glass Cleland and Juanita Brooks, eds., *A Mormon Chronicle: The Diaries of John D. Lee, 1848–1875,* University of Utah Press, Salt Lake City, 1983, V. I, pp. 252–53.

[16] Robert Glass Cleland and Juanita Brooks, eds., *A Mormon Chronicle: The Diaries of John D. Lee, 1848–1875,* University of Utah Press, Salt Lake City, 1983, V. II, pp. 275–76.

[17] Brooks, *John Doyle Lee,* pp. 141–42.

[18] Robert Glass Cleland and Juanita Brooks, eds., *A Mormon Chronicle: The Diaries of John D. Lee, 1848–1875,* University of Utah Press, Salt Lake City, 1983, V. II, pp. 279–80.

[19] Miller, p. 76.

[20] Brooks, *John Doyle Lee,* pp. 288–89.

[21] Robert Glass Cleland and Juanita Brooks, eds., *A Mormon Chronicle: The Diaries of John D. Lee, 1848–1875,* University of Utah Press, Salt Lake City, 1983, V. II, p. 145.

[22] Robert Glass Cleland and Juanita Brooks, eds., *A Mormon Chronicle: The Diaries of John D. Lee,*

1848–1875, University of Utah Press, Salt Lake City, 1983, V. II, pp. 146–47.

[23] Brooks, *John Doyle Lee,* pp. 295–96.

[24] The responsibility or lack of responsibility of the Mormon church and its instruments for the Mountain Meadow massacre is not a subject within the scope of this essay. The information is limited and because of archival practices of the Church of Jesus Christ of the Latter Day Saints not freely available to every interested party. However, a sound examination of what material that has been made public can be found in the scholarship of Juanita Brooks and I direct the interested toward her monograph on the massacre, her biography of John Doyle Lee, and her wonderful editing of Lee's diaries. For what it is worth, I think the details of the tragedy are not as important as the grim inevitability of it, a chilling inevitability that would soon overwhelm the entire nation as it slid into the horrors of the Civil War. The United States in the late 1850s was a culture primed to accept the killing of brother by brother, whether the symbol of the slaughter be named John Doyle Lee, or a bit later appear under the name of John Brown. It was a dark season in our past we would do well to study so that whatever our future differences, we never revisit such bloody and fruitless ground.

[25] Brooks, *John Doyle Lee,* pp. 370–71.

[26] Miller, p. 119.

[27] For a complete account of the now forgotten plan see Elmo R. Richardson, "Federal Park Policy in Utah: The Escalante National Monument Controversy, 1935–1940," Utah Historical Quarterly, V. 33, No. 2, Spring, 1965, pp. 109–33. For intimate look at the extraordinary life of Harold Ickes see T. H. Watkins, *Righteous Pilgrim: The Life And Times of Harold L. Ickes, 1874–1952,* Henry Holt & Company, New York, 1990 (pages 582–83 deal with the effort to set up the Monument which would have been the largest national park unit in the lower forty-eight states). Ickes' career as Secretary of the Interior is basically unbelievable to anyone who has experienced the timidity and scant vision of modern political life. The biography reads as almost a fable and cannot help but be a tonic in these straitjacketed times. Consider this statement by the Secretary in 1937 to a group of road builders: "We are making a great mistake in this generation. We are just repeating the same mistake in a different form that our forefathers have made. Instead of keeping areas … which will add to the wealth, health, comfort and well-being of the people, if we see anything that looks attractive we want to open up speedways through it so the people can enjoy the scenery at 60 miles an hour." (ibid., p. 472)

[28] See *Wilderness At The Edge: A Citizen Proposal To Protect Utah's Canyons and Deserts,* The Utah Wilderness Coalition, P.O. Box 11446, Salt Lake City, 84147, 1900. Pay particular attention to the lucid and passionate introduction by Wallace Stegner.

TECHNICAL INFORMATION

All images were photographed in the 4 x 5 film format using primarily an Arca Swiss FC camera. The lenses used were: 58mm, 75mm, 90mm, 120mm, 150mm, 200mm, 210mm, 300mm, 400mm, 500mm, and 720mm.

Fuji's Velvia and RFP 50 were the films of choice with an 812 warming filter used as needed, to correct color shift during long exposures. The exposure meter was a Pentax Digital Spot. The scenery was by GOD.